3rd International Conference on Agricultural and Biological Sciences (ABS 2017)

IOP Conference Series: Earth and Environmental Science
Volume 77

Qingdao, China
26 - 29 June 2017

Editor:

Mohammad H. Golabi

ISBN: 978-1-5108-4556-5
ISSN: 1755-1307

Printed from e-media with permission by:

Curran Associates, Inc.
57 Morehouse Lane
Red Hook, NY 12571

Some format issues inherent in the e-media version may also appear in this print version.

Printed by Curran Associates, Inc. (2017)

For permission requests, please contact the Institute of Physics
at the address below.

Institute of Physics
Dirac House, Temple Back
Bristol BS1 6BE UK

Phone: 44 1 17 929 7481
Fax: 44 1 17 920 0979

techtracking@iop.org

Additional copies of this publication are available from:

Curran Associates, Inc.
57 Morehouse Lane
Red Hook, NY 12571 USA
Phone: 845-758-0400
Fax: 845-758-2633
Email: curran@proceedings.com
Web: www.proceedings.com

Table of contents

Volume 77

3rd International Conference on Agricultural and Biological Sciences (ABS 2017)
26–29 June 2017, Qingdao, China

Accepted papers received: 17 July 2017
Published online: 25 July 2017

Preface

Papers

IOP Conf. Series: Earth and Environmental Science **77** (2017) 011001 doi:10.1088/1755-1315/77/1/011001

Preface

Thanks to the support and contributions from participants of more than 35 countries, *the 3rd International Conference on Agricultural and Biological Sciences (ABS 2017)* was held successfully from June 26th to June 29th in Qingdao, Shandong, China.

Promoting scientific and technological progress in the field of agricultural and biological sciences has become the consensus of all the countries in the world. Further challenges come from the climate change caused by global warming, environment pollutions, shortage of water resources and various animal and plant diseases pose new requirements for sustainable agricultural development. ABS conference series was held accordingly to provide a platform for scientists and researchers from all over the world to exchange their researches and new findings in those fields.

The previous two ABS Conferences ABS 2015 and ABS 2016 brought together a forum of renowned scientists from the fields of agriculture, horticulture, soil and biology from the leading research institutes and universities from around the world that made them successful international events. Specializing the scope of ABS fields, we have this year interesting studies on animal biology and plant breeding in the field of biological sciences.

More than 300 papers were submitted to ABS2017, and 50 submitted manuscripts have met the scope of IOP Conference Series on Earth and Environmental Sciences to form this special book. After pre-review on the originality and language, peer review process was arranged by the editorial committee. The editorial board was led by the Guest Editor Prof. Mohammad Golabi and reviewers like Dr. Jesus Rodrigo-Comino, Dr, Burak Kocak, Dr. Miguel Sanchez-Monedero etc and 21 best manuscripts were selected after the pre-review and peer review, and those manuscripts provide some insights in the field of soil, environmental and biological sciences.

On behalf of the conference organizers, we would like to take this opportunity to express our sincere thanks to guest editor and all the reviewers for their tremendous efforts and dedication to the conference, to all the authors for their trust and contributions to the conference, as well as all the colleagues from IOP publisher for their support and their endless efforts towards the publication of the proceedings for the Conference. We believe that with their earnest support and contributions, future ABS Conferences would scale new heights.

Below are some interesting pictures from ABS2017:

IOP Conf. Series: Earth and Environmental Science **77** (2017) 011001 doi:10.1088/1755-1315/77/1/011001

Editor

Prof. Mohammad H. Golabi

University of Guam,

Guam-USA

IOP Conf. Series: Earth and Environmental Science **77** (2017) 011002 doi:10.1088/1755-1315/77/1/011002

Peer review statement

All papers published in this volume of *IOP Conference Series: Earth and Environmental Science* have been peer reviewed through processes administered by the proceedings Editors. Reviews were conducted by expert referees to the professional and scientific standards expected of a proceedings journal published by IOP Publishing.

IOP Conf. Series: Earth and Environmental Science **77** (2017) 012001 doi:10.1088/1755-1315/77/1/012001

The comparison of light responses among four species of *Calligonum* L. in early autumn

H Pan[1, 2], Y Feng[1,4] and Y Li[3]

[1] Xinjiang Institute of Ecology and Geography, Chinese Academy of Sciences, Urumqi 830011, China
[2] Chinese Academy of Sciences, Beijing 100049, China
[3] Arid Ecology and Environment Institute, Xinjiang University, Urumqi 830046, China

E-mail: luckfy@ms.xjb.ac.cn

Abstract. The photosynthetic characteristics of 4 species from different *Calligonum* sections were examined in autumn. Light response curves were measured in the field and fitted by the nonrectangular hyperbola model and modified rectangular hyperbola model. Modified rectangular hyperbola model was found greater when the fitting effects of two models were analyzed by relative errors. The comparison of the photosynthesis among four species revealed that *C. pumilum* had the maximum net photosynthetic rate (Pmax) with high utilization of light. The lowest light saturation point (LSP) and apparent quantum yield (AQY) were observed in *C. Calliphysa* which seemed to be the low utilization of light. *C. densum* had the highest photo inhibition coefficient and smallest photo saturation coefficient; in contrast, *C. leucocladum* had the smallest photo inhibition coefficient and highest photo saturation coefficient. Present study revealed that there were differences of photosynthetic characteristics between four species in the genus *Calligonum*. Further studies are needed to elucidate this conclusion with more species and data collected in other season.

1. Introduction

Xinjiang lies in the northwest of China, and is a typical temperate arid zone with a high diversity of landscapes, such as glacier, alpine, basin, plain, grassland and desert. With the uplift of the Qinghai-Tibet Plateau (QTP) and the aridification originated in early Miocene [1] and the following Pleistocene aridification, large scale expansion of sandy deserts developed. The climate changes gave rise to a diversified ecosystems and desert flora. However, the amount and diversity of plant species decreased due to the enhanced aridification, thus an urgent vegetation protection and ecological restoration is required.

The genus *Calligonum* contains 35 species in the world, among which 22 were found in Xinjiang.Based on fruit morphological characteristics, *Calligonum* has been divided into four sections, including *sect. Calliphysa* (Fisch. et Mey) Borszcz, *sect. Pterococcus* (Pall.) Borszcz, *sect. Calligonum*, and *sect. Modusa* Sosk et Alexandr. According to the Flora Xinjiangensis, there is 1 species in *sect. Calliphysa*, 3 in *sect. Pterococcus*, 5 in *sect. Calligonum* and 13 in *sect. Modusain* [2]. *Calligonum* species are drought-tolerant, and mainly distributed on semi mobile or stable sand dunes and sandy lands in Xinjiang, playing an important role in wind breaking and sand fixation.

IOP Conf. Series: Earth and Environmental Science **77** (2017) 012001 doi:10.1088/1755-1315/77/1/012001

The photosynthesis is important in plants nourishment and growth. Photo response characteristics is the most important tool to study plant photosynthesis [3,4]. Several studies concerning the photosynthesis of *Calligonum* have been published, dealing mainly with the photosynthetic physiology and drought resistance of different species of *Calligonum* in different areas [5-7]. However, no studies have focused on the comparison of photosynthesis among species from different sections of *Calligonum*, and it may shed new light on our knowledge of *Calligonum* species' divergence and adaptation. In addition, some species of *Calligonum* are re-blooming in early autumn, so more attention should be paid to the photosynthesis in this season. In the present study, the light response curves of four species of *Calligonum* were measured and analyzed by the nonrectangular hyperbola model [8] and modified rectangular hyperbola model [9]. The aim of this study was to provide further knowledge of the photosynthetic physiological characteristics of *Calligonum* and make a contribution to their protection and restoration.

2. Materials and methods

2.1. Plant materials

The study area was in Turpan desert botanical garden of Chinese Academy of Sciences Institute of Xinjiang Ecology and Geography (Turpan, 42°91'N, 89°19'E). 4 species from 4 *Calligonum* sections were examined, (1) *C. leucocladum* from *sect. Pterococcus*; (2) *C. pumilum* from *sect. Modusa*; (3) *C. calliphysa* from *sect. Calliphysa*; (4) *C. densum* from *sect. Calligonum.* Three individuals were chosen to represent the average growth of each species, and fully developed, mature, and healthy shoots were selected and carefully tested.

2.2. Photosynthetic analysis and Light response measurements

The measurements were taken under continuous cloudless weather at 11:00 to 13:00. The photosynthetic active radiation (PAR) gradient was set at 0, 25, 50, 100, 150, 200, 300, 500, 800, 1000, 1200, 1500, 1800 and 2000 μmol $m^{-2}\cdot s^{-1}$. The physiological parameters were automatically recorded by LI-6400 portable photosynthesis system.

Light response curves were created with two models, nonrectangular hyperbola model [8] and modified rectangular hyperbola model [9].

2.3. Data analysis

Statistical analyses were carried out using the Statistical Product and Service Solutions (SPSS19.0) software. Figures were drawn with Origin 9.0.

3. Results

As shown in figure 1, net photosynthetic rates (Pn) of the 4 species exhibiting similar S-shape. The Pn values increased along with increase of PAR in low light intensities and reached a plateau (maximum net photosynthetic rate, Pmax) when PAR reached 1200–1500μmol $m^{-2}\cdot s^{-1}$. All the 4 species have similar light saturation point (LSP). When PAR was higher than LSP, 4 species maintain a high level of Pn, with significant differences ($p < 0.01$).

The physiological parameters, such as light compensation point (LCP), light saturation point (LSP), maximum net photosynthetic rate (Pmax) and dark respiration rate (Rd) were obtained by the nonrectangular hyperbola model and modified rectangular hyperbola model. These parameters provide a basis to evaluate the photosynthesis systems, and study the photosynthesis ability and light adaptability of plants. Table 1 shows the parameters measured and fitted by two models, and the apparent quantum yield (AQY) were estimated by applying linear regression under low-light intensities (0-150 μmol· $m^{-2}\cdot s^{-1}$). The light response curves measured and predicted by the two models were shown in figure 2.

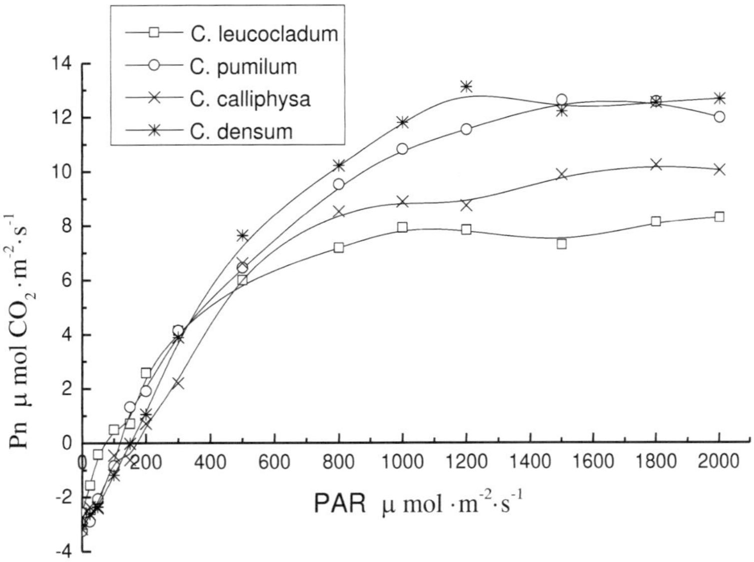

Figure 1. The photosynthesis light response curves of 4 species (4 sections).
PAR: photosynthetic active radiation; Pn: net photosynthetic rate.

Table 1. The photosynthetic parameters of 4 species (four sections) measured and fitted by two models.

Groups	Species	Model	Pmax	LSP	LCP	Rd	AQY/α	Adjusted R^2
Sect. Pterococcus	*C. leucocladum*	Measured	8.3	1600	80	2.56	0.022	
		Model I	10.736		89	2.209	0.026	0.993
		Model II	8.09	1588	80	2.476	0.037	0.995
Sect. Modusa	*C. pumilum*	Measured	12.62	1700	120	3.24	0.03	
		Model I	17.974		122	3.405	0.03	0.996
		Model II	12.39	1755	120	2.268	0.033	0.999
Sect. Calliphysa	*C. calliphysa*	Measured	10.05	1600	140	2.73	0.017	
		Model I	13.254		153	3.02	0.02	0.993
		Model II	10.13	1643	141	3.237	0.026	0.994
Sect. Calligonum	*C. densum*	Measured	13.12	1500	130	3.17	0.021	
		Model I	16.689		144	3.403	0.024	0.997
		Model II	13.04	1525	133	3.664	0.03	0.998

Model I : the nonrectangular hyperbola model; Model II : modified rectangular hyperbola model; Pmax (μ mol CO_2 m^{-2} s^{-1}): light-saturated net photosynthetic rate; LSP (μ mol $m^{-2}\cdot s^{-1}$): the light saturation point; LCP (μ mol m^{-2} s^{-1}): light compensation point; Rd (μ mol $m^{-2}\cdot s^{-1}$): dark respiration rate; AQY: apparent quantum yield estimated by applying linear regression; α: initial quantum efficiency or intrinsic quantum yield estimated by two models; Adjusted R^2: the conformity coefficient of models.

Though both models had high determination coefficients (R^2>0.99), the fitting effects presented significant differences between two models (figure 2), the fitting results of Pmax with the nonrectangular hyperbola model were significantly higher than the measured values ($p < 0.01$), the relative errors (RE) between the measured and fitted values was 27%-42%. However, modified rectangular hyperbola model fitted the light response curves of *Calligonum* well and the RE is 1%-

IOP Conf. Series: Earth and Environmental Science **77** (2017) 012001 doi:10.1088/1755-1315/77/1/012001

2.5%. The fitting results for LCP with two models were largely identical to the measured values (p>0.05), the RE of the nonrectangular hyperbola model and modified rectangular hyperbola model was 2%-11% and 0-8%, respectively. The rectangular hyperbolic model is unable to estimate LSP on account of the model is an asymptotically saturating curve without a clear maximum within the range of the data [10-12], while the fitting values of LSP were the most similar to the measured values by modified rectangular hyperbola model (RE is 0-6%).

Pmax fitted by modified rectangular hyperbola model was that *C. pumilum* and *C. densum* was significant higher than other two species (p<0.01), *C. leucocladum* is the lowest and show significant differences (p<0.05). The comparison of LSP values was that *C. pumilum* > *C. leucocladum* > *C. densum*> *C. calliphysa*. The values of LCP were that *C. calliphysa*> *C. densum*> *C. pumilum* > *C. leucocladum.* The β values by modified rectangular hyperbola model were *C. densum*> *C. calliphysa* > *C. pumilum* > *C. leucocladum.* The γ values were that *C. leucocladum* is significantly higher than other species (p<0.01), and others show no significant differences (p>0.05). The AQY values showed that *C. pumilum* > *C. Densum* > *C. leucocladum* > *C. Calliphysa* (p<0.01).

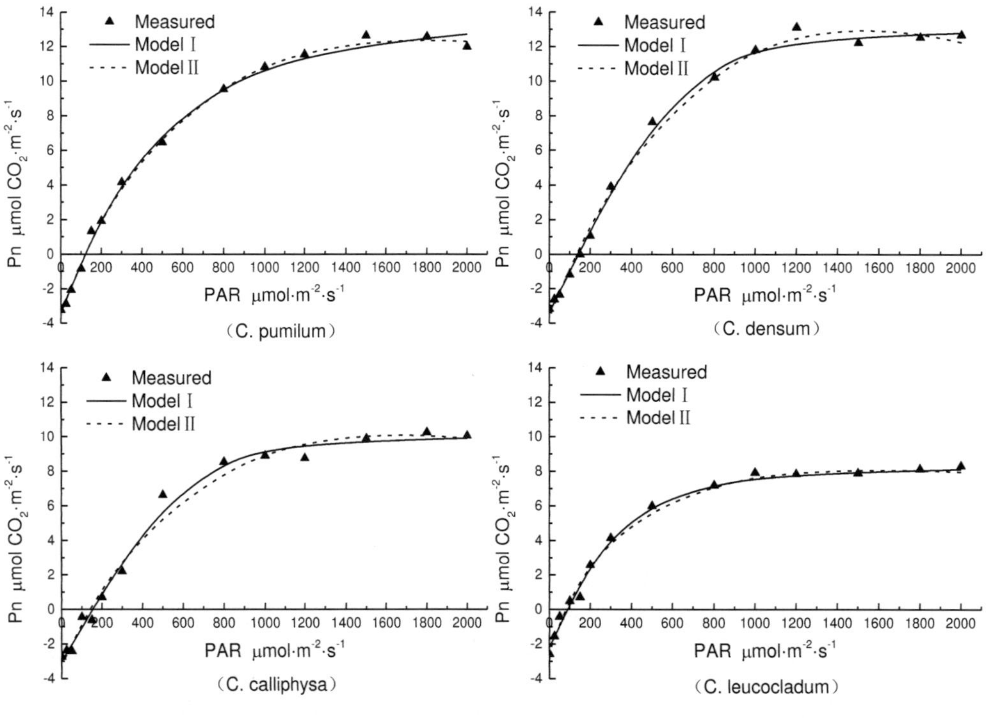

Figure 2. The light response curves of 4 species(4 sections) measured and predicted by two models. PAR: photosynthetic active radiation; Pn: net photosynthetic rate; Model Ⅰ: the nonrectangular hyperbola model; Model Ⅱ: modified rectangular hyperbola model.

4. Discussion

4.1. The comparison and fitting of two models

Light response curve models were widely used in plant photosynthetic physiological studies, among which nonrectangular hyperbola and modified rectangular hyperbola model were the most commonly adopted models. The fitting effects of light-response models were variable among different plants, environments and seasons [13, 14]. The nonrectangular hyperbola model was also an asymptote same to the rectangular hyperbola model, which couldn't estimate the values of LSP accurately [15]. Several studies indicated that the nonrectangular hyperbola model couldn't fit photo inhibition plants because this model had not extreme values and usually overestimated Pmax when the Pn was declining after LSP [12]. Modified rectangular hyperbola model introduced the photo-inhibition term β and the photo-saturation term γ, which could estimate the value of LSP accurately because it wasn't an asymptotically saturating curve. This model also had high accuracy and applicability in fitting the process that curve is declining or remaining consistent after LSP.

The present research is again proves that the nonrectangular hyperbola model significantly overestimates the values of Pmax and modified rectangular hyperbola model is mostly consistent with measurements. Although both models had higher determination coefficients (R^2>0.99), R^2 didn't indicate that the model is consistent to the actual measurement. RE (relative errors) should be a new parameter for the quantitative evaluation of the fitting effect of light response models [16], the less the value of RE is, the more perfect of the model is. The comparison of the RE between two models show that modified rectangular hyperbola model is the most suitable light-response model for *Calligonum*. However, it's hard to say that modified rectangular hyperbola model is the best model for *Calligonum* under any conditions. It is difficult to compare the parameters because they were of different meanings in two models. The value of AQY resulted in linear regression were not precise due to the different ranges of light intensity or different numbers of data points. In addition, the fitting effects of modified rectangular hyperbola model might be related to plant type, different temperatures or different carbon dioxide and soil water conditions [12,14,17]. Thus, further researches are needed to demonstrate the fitting effects of *Calligonum* by modified rectangular hyperbola model.

4.2. The differences among 4 sections of Calligonum

Photosynthesis plays an important role in plants growth because it provides organic matters for their development. Plants adjust their ecological strategy to improve the efficiency of photosynthesis under different habitats, for example, allocating more biomass into above ground part and changing root/shoot ration to compete for light resource [18], and controlling the stomata conductance to adapt to drought stress [19]. *Calligonum* species with higher ability of photosynthetic efficiency are superior to other species for accumulating organic materials. Furthermore, high ability of photosynthesis contributes to high resistance and adaptability to variety of environments. The lower the AQY or the lower the LCP, the stronger ability of the plants to use weak light, it means the organic matter of plants start positive growth under very low PAR. What's more, these plants have high light use efficiency under the light limited conditions, thus increasing the accumulation of organic material. Light ecological amplitude to plants is the range between LCP and LSP, and it indicates the ability of plants to different illumination intensity. A higher value of LSP and Pmax shows a high utilization efficiency of high PAR. In this study, the photosynthetic parameters of four *Calligonum* species in the same habitat were analyzed by linear regression and modified rectangular hyperbola model.

We found that *C. pumilum* had the most widely light ecological amplitude, the highest light-saturated net photosynthetic rate (Pmax) and light utilization efficiency; *C. pumilum* had not only strong drought resistance but also favorable adaptability to high light quantum flux density that may be why they are widely distributed and usually the dominant or primary constructive species in their ecological communities. Several species in *sect. Modusa* are endemic that evolved in the severely arid environment in Xinjiang, such as *C. taklimakanense*, *C. roborovskii*, *C. kurlaense*, *C. yengisaricum*, and *C. juochiangense* [2]. *C. calliphysa* have narrow light ecological amplitude and the lowest light utilization efficiency. As a result, the habitats of *C. calliphysa* is poor under natural conditions, and they are usually prostrate on the ground seems to be dying. *C. calliphysa* seldom is the dominant species or constructive species. The significant differences between *C. pumilum* and *C. calliphysa* may due to the physiological adaptability and resistance of different sections. Li [20] suggested that the fruit of *Calligonum* grows to be shallow trench and multi-spine density to adapt to the complex and harsh environment. Lv [21] suggested that *C. calliphysa* is the oldest section and *C. pumilum* is the youngest section after genetic variation and population evolution. The theory of evolution and survival of the fittest all prove that *C. calliphysa* has the weakest photosynthetic capacity while *C. pumilum* has the strongest photosynthetic capacity.

In modified rectangular hyperbola model, β is the correction factor and considered as photo-inhibition term, γ is the coefficient and considered as photo-saturation term, their biological significances are the product of cross section of photosynthetic pigment in PS II for quantum absorption and average life of photosynthetic pigment in the excited state [22]. The significance suggests that the plants are easily photo-inhibited if the value of β is large, while plants are easily photo-saturated if the value of γ is large. The light use efficiency of plants depends on the comprehensive of β and γ, the higher the β and γ, the low use efficiency of plants. In this study, specie of *C. densum* has the highest value of β and the lowest value of γ, it may reveal that *C. densum* is most easily affected by photo inhibition while most hardly reaching photo saturation, while *C. leucocladum* behaves like a shade plant and is most hardly affected by photo inhibition and most easily reaching photo saturation. In fact, we didn't find the similar trend in the light response curves, and this is because of the neutralization combination of the two parameters. The lower the β and γ is, high utilization of light of the plants is. This confirms the theory that *C. densum* and *C. leucocladum* are the transition stage of evolutionary development and have the medium utilization of light. *Calligonum* as an important windbreak and sand-fixing shrub is usually introduced in shelter forest construction and ecological restoration. *C. densum* and *C. leucocladum* may both not suitable for introducing as dominant or constructive species in saline, extreme drought conditions, because they are easily affected by environmental factors. Furthermore, the photo inhibition, photo saturation and low values of Pmax lead to decline of light utilization rates and photosynthetic rates, which resulting in low amount of organic accumulation and poor growth under severe interferences.

Although it was advocated that photosynthesis of plants is not as good as that of in summer [19], the present study still provides some meaningful implications that photosynthesis differences are existing among the four species and each of them have their own characteristics. Studies have illustrated that the various phenology of *Calligonum* show different responses of plants to the same environment associated with seeds morphology [23]. Kang [24] indicated that the intersection differences of *Calligonum* under same habit decided by the heredity of fruit differences. Our study showed the different *Calligonum* sections have different microenvironments, and their various physiological adaptability and resistance lead to different utilization degrees of illumination. However, further evidences to confirm the differences of light responses among *Calligonum* sections are needed. Next year, the photosynthesis of more *Calligonum* species from four sections will be studied. Moreover, anatomy and carbon isotope will also be included in our future research to gain a deeper insight of the divergence among *Calligonum* species, as well as the evolution and adaption of the genus in arid environment.

Acknowledgments
This research was supported by National Science Foundation of China "Studies on kinship, geographical replacement law and evolution reason of *Calligonum* plants in Xinjiang" (NSF-41271069) and the Province Science Foundation of Xinjiang (Y55315100).

References
[1] Guo Z T *et al.* 2002 Onset of Asian desertification by 22 Myr ago inferred from loess deposits in China *Nature* **416** 159-63
[2] Mao Z M 1993 *Calligonum* L. *Flora Xinjiangensis* vol 1 (Urumqi: Xinjiang Science and Health Press) pp 267-74
[3] Chen Z Y, Peng Z S, Yang J, Chen W Y and Ou-Yang Z M 2011 A mathematical model for describing light-response curves in *Nicotiana tabacum* L. *Photosynthetica* **49** 467-71
[4] Sakai T, Yamamoto S S and Akiyama T 2005 Microsite variation in light availability and photosynthesis in a cool-temperate deciduous broadleaf forest in central Japan *Ecol. Res.* **20** 537-45
[5] Yan H L, Zhang X M, Xu H and Yao S J 2007 Responses of *Calligonum arborescens* Photosynthesis to Water Stress in Tarim Highway Shelterbelt *J. Desert Res.* **27** 460-5

[6] Zhu C G, Chen Y N, Li W H and Ma X D 2014 Effect of herbivory on the growth and photosynthesis of replanted *Calligonum caput-medusae* saplings in an infertile arid desert *Plant Ecol.* **215** 155-67
[7] Zhu J T, Li X Y, Zhang X M, Lin L S and Yang S G 2011 Photosynthetically and ecophysilogical characterstics of *Calligonum roborwasikii* in different altitudes on the northern slope of Kunlun Mountain *Acta Ecol. Sin.* **31** 611-9
[8] Marshall B and Biscoe P V 1980 A model for C3 leaves describing the dependence of net photosynthesis on irradiance *J. Exp. Bot.* **120** 29-39
[9] Ye Z P 2007 A new model for relationship between irradiance and the rate of photosynthesis in *Oryza sativa Photosynthetica* **45** 637-40
[10] Ye Z P and Yu Q 2007 Comparison of a new model of light response of photosynthesis with traditional model *J. Shenyang Agri. Uni.* **38** 771-5
[11] Ye Z P and Li J S 2010 Comparative investigation light response of photosynthesis on non-rectangular hyperbola model and modified model of rectangular hyperbola *J. Jinggangshan Uni.* (Natural Science) **31** 38-44
[12] Lang Y, Wang M, Zhang G C and Zhao Q K 2013 Experimental and simulated light responses of photosynthesis in leaves of three tree species under different soil water conditions *Photosynthetica* **25** 370-8
[13] Xia J B, Zhang G C, Wang R R and Zhang S Y 2011 Effect of soil water availability on photosynthesis in *Ziziphus jujube* var. *Spinosus* in a sand habitat formed from seashells: Comparison of four models *Photosynathetica* **52** 253-61
[14] Xia J B, Zhang S Y and Zhang G C 2013 Critical responses of photosynthetic efficiency in *Campsis radicans* (L.) Seem to soil water and light intensities *Afr. J. Biotechnol.* **10** 17748-54
[15] Wang R R, Xia J B, Yang J H, Zhao Y Y, Liu J T and Sun J K 2013 Comparison of light response models of photosynthesis in leaves of *Periploca sepium* under drought stress in sand habitat formed from seashells *Chinese J. Plant Ecol.* **37** 111-21
[16] Yun Y L and Cao B 2007 The data fitting based on the optimized relative error *Stat. Decis.* **21** 15-6
[17] Yan X H, Yin J H, Duan S H, Zhou B, Wen-Hai H U and Liu S 2013 Photosynthesis light response curves of four rice varieties and model fitting *Chinese J. Ecol.* **32** 604-610
[18] Song C, Zeng F J, Liu B, Zhang L G, Luo W C, Peng S L and Stefan K A 2012 Influence of water condition on morphological characteristics and biomass of *Calligonum caput-medusae* Schrenk seedlings *Chinese J. Ecol.* **31** 2225-33
[19] Su P X, Zhao A, Zhang L, Du M and Chen H 2003 Characteristic in photosynthesis, transpiration and water use efficiency of *Haloxylon ammodendron* and *Calligonum mongolicum* of desert species *Acta Bot Boreal-Occident Sin* **23** 11-7
[20] Li D L, Wang J H and Ma Q L 2009 Numerical Taxonomy of Fruit Characteristic of *Calligonum* L *J. Desert Res.* **29** 718-22
[21] Lv W D 2008 Studies on Classification of *Calligonum* L. (Gansu Province: Lanzhou University)
[22] Ye Z P and Kang H J 2012 Study on biological significance of coefficients in modified model of photosynthesis irradiance *J. Yangzhou Uni.* (Agricultural and Life Science Edition) **33** 51-7
[23] Zhao Y F, Shi W, Pan B R, Yi L and Liu R 2014 The phenophase response of *Calligonum* L. in Turpan eremophytes botanical garden to temperature change in recent 29 years *J. Desert Res.* **34** 732-9
[24] Kang X S, Pan B R, Duan S M, Suo F Y, Zhang Y Z and Shi W 2012 Comparison of flowering phenology and reproductive traits of four species of *Calligonum* L. (Polygonaceae) under Ex-situ conservation at the Turpan eremophytes botanical garden in Xinjiang, China *J. Desert Res.* **32** 1315-27

The exploit of cereal embryo structure for productive reasons by *in vitro* techniques

C Savaskan[1]

Faculty of Arts and Sciences, Department of Biology, Süleyman Demirel University, Isparta, Turkey

E-mail: savaskancd@gmail.com

Abstact. There are two main sides of our works exploiting embryo structure in durum wheat and some other cereals. First is haploid (or doubled haploid) embryo production using anther or microspore culture or intergeneric crosses, to ameliorate desirable characters genetically homozygote. Secondly, to develope convenient embryo culture technique in order to be stored and cultivated longer time of genotypes without being alien pollination etc. in field conditions. For that reason, two different auxin and also their combination with kinetin were used for mature embryos of wheat genotypes (hexaploid and tetraploid), to understand efficient dose for calli production and plant regeneration in plant tissue culture. Modified MS media were used adding a single dose of arabinogalactan protein (AGP) and without adding for regeneration. In further step of this study, most efficient auxin+kinetin combination which is determined previous research, it was used in the same modified MS medium to produce calli production and plant regeneration in three different genotypes (hexaploid and tetraploid wheat and diploid barley). Data were calculated in five different developmental stages of treatments. All statistical analysis of data were performed and means were compared with Duncan's test. Genetics and morphological effects of AGP on genotypes were discussed with the results of variance analysis. Simple correlation coefficient (r) was calculated base on the main values of replications.

1. Introduction

An embryo is an useful organisation to manipulate with various techniques for agricultural or biological objectives particularly in angiosperms. It can give sufficient answers in different developmental stages and parts which are related with either genetical components or anatomical structures. Monocotyl plants (mainly cereals) carry a tremendous capacity for showing of these maipulations in further generations, such as variation in genetics, differences in anatomy or other desirable characters in agriculture, because of their numerical abundancy in morphology. A mature embryo as a part of a seed besides endosperm and seed coat creates fertile individual for further generation of its species (figure 1).

A seed is a very valuable source that can be belonging to a wild type, landrace or a cultivar improved by selection with breeding methods. Adaptation, yield and quality or related values are features expressed by gene(s) which are found as genetic materials into seed. The long term usage of these features is necessary for humankind in production and nutrition. However, its productivity reduces with many uses, enviromental factors or foreign pollinations etc., during the cultivation. There are two main sides of our works generally.

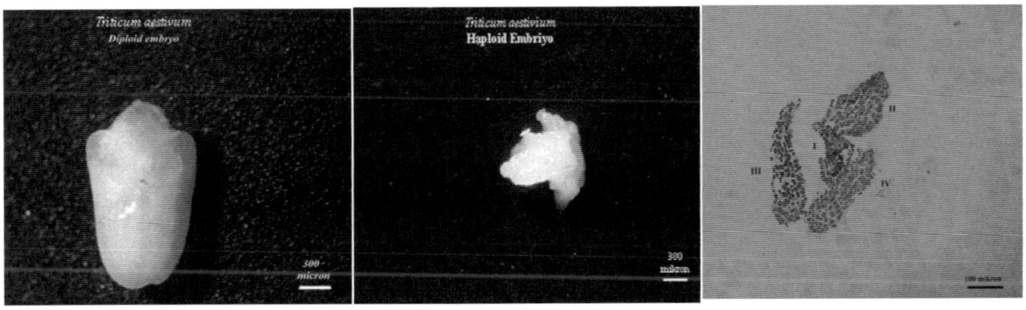

Figure 1. A mature diploid bread wheat embriyo (*Triticum aestivum* L.,) (left). A haploid embriyo produced using wheat x maize crosses (middle) and its anatomical structure (right) [1]. I. radicle and coleorhizae, II. plumula and leaf primordium, III. scutellum, IV. stem (at right).

- haploid (or doubled haploid) embryo production using anther or microspore culture or intergenetic crosses, to ameliorate desirable characters as genetically homozygote to long term use,
- to exploit of mature embryo structure for protection of phenotypic characters without losing genes in genotypes. Thus, ameliorated genotypes can be stored or used in agronomy for longer time.

Here, the point of view is to contribute to own values of nature for the environment rather than endustrial approach, in other words, the prevail subject is the protection of varieties for the beneficial of small size farming, using tissue culture technique facilities. Because there are many varieties improved by genetical and breeding methods, particularly in crop species all over the world, up to now. These genetical sources, landraces and old or obselet cultivars can be used again for further generations without sharing huge natural resources and budgets for new varieties (figure 2).

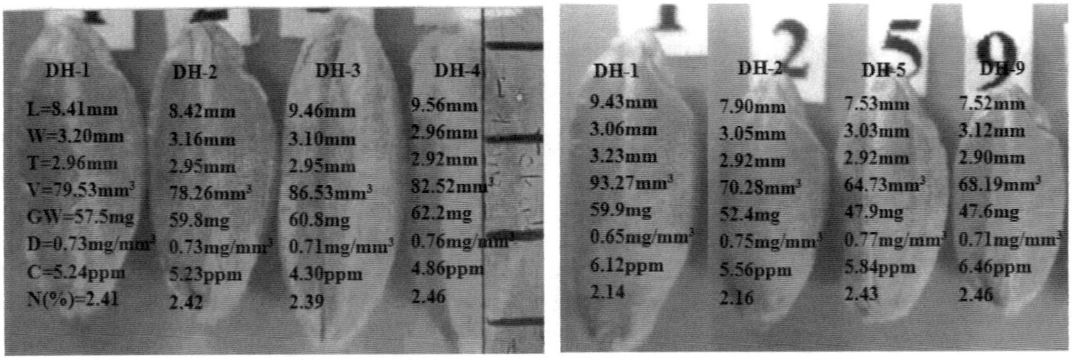

Figure 2. Revealing of variation as different genotypes from old/obselet durum wheat landraces after making homozygote of alleles by doubling of chromosomes using intergeneric crosses or embryo rescue. Grains of doubled haploid (DH) genotypes derived from *T.durum* cv. Berkmen 469 (left) and Kunduru 1149 (right) [4].

In tissue culture techniques, the main function is to product abundant and healty productive embryos or embryoid calli in induction and differentiation for regeneration on culture media, respectively. Therefore supportive growth regulators, proteins or other physical factors will be able to provide a rich culture medium for plant materials (explants). For that reason, we considered two important details in mature embryo studies using two ploidy levels of wheat and also a diploid barley. As first, the effectiveness of growth regulators (auxin and cytokinin) for induction embryoid-calli was searched in wheat [2]. Secondly, the effects of various amounts of Arabinogalactan-Proteins (AGPs), which are described extracellular proteoglicans in most cases, were used for high regeneration capacity in hexaploid and tetraploid wheat and also a diploid barley genotypes [3]. In this presentation the importance of genetic and morphological effects of AGPs in differentiation were discussed in various ploidy levels of these genotypes, exploiting mature embryo structure.

2. Materials and methods

2.1. Plant materials

In our researches, mainly durum wheat has been used as plant material. Wheat is one of the important crop plant economically and has a long period of evolutionary history. There are two very important species widely used by humankind, *Triticum aestivum* (hexaploid, bread wheat) and *T. durum* (tetraploid, durum wheat). Besides these, some other cereals have been worked in our tissue culture studies in anther culture tecnique and intergeneric crosses for germplasm evaluation and for also comparison of haploid-diploid embryo structure [1,4,5].

In mature embryo culture studies, three different ploidy levels of cereals, bread wheat (hexaploid-AABBDD), *T. aestivum* L. *cv.* Ikizce-96 (2n=6x=42), durum wheat (tetraploid-AABB) *T.durum* Desf. *cv.* Mirzabey (2n=4x=28), and barley (diploid-AB) *H.vulgare* L. *cv.* Tokak 157/37 (2n=2x=14) were chosen between Anatolian cultivars. Seeds of cultivars were obtained from the Field Crops Central Research Institute, Ministry of Agriculture and Rural Affairs in Ankara, Turkey [2,3].

2.2. Hormones and Arabinogalactan Proteins (AGPs)

Hormones: Four different hormone combinations were treated by adding to MS (modified) medium [6] supplemented with 20 g L^{-1} sucrose and 7 g L^{-1} agar, to understand the effects of calli production in each of treatment groups. Auxins (Dicamba and 2,4-D) 5mg L^{-1} and also their combinations with cytokinin (Kinetin) 1 mg L^{-1} were prepared to searched in the first treatment group and among them, Dicamba+Kinetin (5 mgL^{-1} +1 mgL^{-1}) hormones were given the best result [2]. That determined combination was used for calli production in the second step of study [3].

AGPs: AGPs have been shown to be expressed throughout the plant kingdom and have been considered to have important roles in plant growth and developments. The majority of the AGP has a protein content less than 10%, and contains more than 90% carbohydrate. They are water soluble non-starch pentosan group glycoproteins and can be found many parts of a plant, in vacuoles [7] particularly cell-wall structure [8]. Wilkinson *et al* (2013) reported that numerous AGP sequence types were identified in Poaceae family members in *Agropyron mongolicum, Secale cereale, Oryza sativa* subsp. *japonica* and *Sorghum bicolor*. Since the AGPs are rich in hydroxyproline they are classified as hydroxyproline-rich glycoproteins (HRGPs) [9]. Coskun *et al* (2010) found that AGP increased plant regeneration in wheat genotypes when they were added 10 ml L^{-1} into MS medium [2]. However, at the following study, they determined that 5 mg L^{-1} of AGP is the best quantitiy for a suitable regeneration system between six different AGP concentrations (0,2,5,7,10,12 mg L^{-1}), in three ploidy levels of cereals including mature embryos of bread and durum wheat and diploid barley [3].

2.3. Data analysis

Data were analysed as follows: Callus induction efficiency = (Number of calli/Number of mature embryo) x 100. Plant regeneration efficiency =(Number of plant regenerant/Number of mature embryo) x100. Culture efficiency = (Callus induction efficiency/Plant regeneration efficiency) x 100. Weight of callus = Each callus was weighed on their 14th day of culture. Measurement of callus = Each callus was measured on their 14th day of culture. The length of root, stem and total plant heigth= Each plant was measured after 6 weeks of culture.

Statistics: Each of petri plate containing 20 embryos was considered as single of replication and treatment was conducted according to randomised complete design with four replications for each AGP regeneration treatment. SPSS 15.0 software programme was used for istatistical analysis and means were compared with Duncan's Test in each means of application of hormones and means between culture media groups. Simple corelation (r) were calculated based on the mean values of replications [10].

3. Results and discussion

The results of the influential effects of some of hormone treatments on callus production derived from mature wheat embryos, and, efficient plant regeneration with AGPs on various ploidy levels of cereals have been discussed in Coskun *et al* (2010, 2013) [2, 3]. Analysed data (parameters) were performed with variance analysis and means were compared with Duncan's test in each of two researches.

Callus weight was statistically different between genotypes (P<0.01) (table 1). The lowest weight one was found in Ikizce-96 (79.91 mg) while its callus induction (93.75%) was the highest (table 1). Tokak 157/37 with the smalest number of genome (AB) produced the weightiest (179.95 mg) and largest callus (6.18 mm) (table 1). In contrast diploid Tokak 157/37 produced less regenerant plants than other genotypes (tetraploid and hexaploid) [3]. Savaskan *et al* (1999) used Tokak 157/37 in androgenesis to produce haploid regenerant plants, and they observed that this genotype produced less regenerant plants than other diploid barley genotypes (Cumhuriyet-50, Anadolu and Obruk) although it produced much more embryoid calli than the others [5]. Bi and Wang (2008) compared diploid and tetraploid wheat genotypes for capacity of totipotency and mentioned that tetraploid wheats produced more calli and regenerant plants than diploids [11]. Callus induction didn't show differences statistically between either two hexaploid and tetraploid wheat genotypes [2] or wheats and diploid barley [3] when Dicamba +Kinetin (5 mg L^{-1}+ 1 mg L^{-1}) combination was used in MS induction medium.

Table 1. Variance analysis and means of calli procedures of three genotypes.

Genotype	Genome	Callus induction (%)	Weight of callus (mg)	Diameter of callus (mm)
İkizce-96	AABBDD	93.75 ± 3.53 a	79.91 ± 3.23 a	4.56 ± 0.25 a
Mirzabey	AABB	93.50 ± 3.57 a	83.84 ± 1.79 b	4.69 ± 0.11 a
Tokak 157/37	AB	91.25 ± 2.31 a	179.95 ± 3.13 c	6.18 ± 0.25 b
F-value		n.s.	**	**

*Entries within column followed by the same letter are not significantly different (P<0.05).
n.s. =non significant, * P<0.05, **P< 0.01, [3].

3.1. Genetic effects of AGPs on differentiation

Researchers stated that the wheat Gsp-1 gene resulted in reduced contents both AGP and grain softness protein (GSP-1) in mature wheat grains confirming that these components are encoded by the same gene [7,12]. In earlier studies, other researchers had already reported that Gsp-1 genes are localised on the 5A1, 5B1 and 5DS chromosomes of wheat and linked to the grain hardness lokus (<I>Ha</I>) on 5D [13,14]. In that point, it is possible to understand that the degreasing of gene effects on calli differentiation according to the genom numbers of genotypes were found significantly different (table 2). In other words, differences between regeneration capacity (%) and also culture efficiency (%) of genotypes were important (P<0.01). Similarly the variance for AGP treatments for these two parameters were found important as (P<0.05) and (P<0.01) respectively (table 2). Genotype x AGP interactions on three genotypes (ploidy levels) were found significant (P<0.01) at both parameters related with differentiation, regeneration capacity (%) and culture efficiency (%) (table 2).

3.2. Morphogenesis of AGPs for plant development

Also hexaploid genotype (Ikizce-96) gave the highes value for root and stem length and total plant heigth as 12.46, 19.00 and 31.46 cm in plant regeneration using MS+5 mg L^{-1} AGP while diploid Tokak 157/37 produced the lowest values in the same group of regeneration medium, 10.39, 13.39 and 23.78 cm, respectively [3]. The length of root and stem and also total plant heigth were measured in all regenerant plants after six weeks [3]. In genotypes, morphological values of regenerant plants (root and stem length and total plant heigth) improved in existance of AGP (P<0.01) (table 2). At the organ level, AGPs are found in leaves, stems roots, floral parts and seeds [15,16]. For that reason, it is possible to expected that genotype x AGP interaction in morphogenesis (plant growth and development) was found significantly different (table 2).

Table 2. Analysis of variance for AGP treatments on three different genotypes.

		F-Value				
Variation Sources	df	Root Length (cm)	Stem Length (cm)	Plant Height (cm)	Regeneration Capacity (%)	Culture Efficiency (%)
Genotype	2	8.196**	11.275**	13.269**	21.973**	16.842**
AGP	5	5.223**	3.935**	4.286**	3.584*	4.869**
Gen. x AGP	10	27.087**	28.457**	29.895**	66.768**	67.566**

All characters correlated to each other on positive side when AGP added 5 mg L^{-1} to MS medium for embryo culture and simple correlation between culture efficiency and regeneration capacity was found important ($P<0.01$) (table 3). Also, either culture efficiency or regeneration capacity were correlated with root length significantly important ($P<0.01$). Correlation between culture efficiency and root length was found as r = 0.9999** (table 3) and the effect of auxin (Dicamba) can be resulted in to this value. Other important auxins naphtol acetic acid (NAA) and indol acetic acid (IAA) had been used in a previous barley anther culture technique for remedy to rootlessness at the regenerant plants which is an important problem for anther culture [5]. In that study, high amount of NAA or IAA (10 mg L^{-1}) had been added to regeneration medium supplemented with 4 mgL^{-1} charcoal to promote rooting [5]. On the other hand, culture efficiency and regeneration capacity didn't correlate with stem length importantly (table 3). The physiological events of hormones and proteins might have been prevailed more than genetical effects in stem development.

Table 3. Simple correlation coefficients (r) between the parameters (characters) of three ploidy levels of genotypes according to the results of 5 mg L^{-1} AGP applications.

Parameters of treatments	r
Culture efficiency – regeneration capacity	0.9988**
Culture efficiency - root length	0.9999**
Culture efficiency – stem length	$0.9500^{n.s.}$
Stem length - regeneration capacity	$0.9646^{n.s.}$
Root length – regeneration capacity	0.9988**

Significant at the 0.01**,0.05* probability levels, non significant$^{n.s.}$

4. Conclusion

- A mature embryo is a vital part of dry dormant seed in plants. Biological and biotechnical manipulations in cereal embryos are being promising procedures for many targets related with basic researches and agronomy [1,4] .
- Dicamba is a stronger auxin in wheats particularly combination with Kinetin than 2,4-D. Dicamba+Kinetin (5 mg L^{-1} + 1 mg L^{-1}) enriched the induction medium and supported calli production significantly [2].
- In various ploidy levels of genotypes, modified MS with 5 mg L^{-1} AGP was given best results in all parameters related with regeneration and morphogenesis. Also, 2 mg L^{-1} of AGP to regeneration medium was provided succesfull results [3]. It may be an economical amount for larger practises and demonstrations.
- According to the results, AGP can be used as a supportive agent in mature embryo culture in cereals. Because it intreacted with genotypes in respect of increasing regeneration and morpgogenesis significantly ($P<0.01$) (table 2). However, correlation between regeneration

capacity with stem length didn't show important relationship although it correlated with root length significantly (table 3). This can be interpreted that genetic effects of AGP changed in different parts of plants.

References

[1] Ozkara A and Savaskan C 2014 Wheat maize crosses for haploid embryo production and comparison of haploid-diploid embryo structure *Res. J. Biotech.* **9** 32-7
[2] Coskun Y, Duran R E and Savaskan C 2011 Influential effects of arabinogalactan-proteins on plant regeneration using calli derived from wheat mature embryos *Afric. J. Agric. Res.* **5** 2439-45
[3] Coskun Y, Duran R E, Savaskan C, Demirci T and Hakan M T 2013 Efficient plant regeneration with arabinogalactan-proteins on various ploidy levels of cereals *J. Integr. Agric.* **12** 420-5
[4] Savaskan C, Akıncı C, Donmez E, Keser M and Yalvac K 2003 The relationships between the grain characters of Anatolian durum wheat genotypes *10th Inte. Wheat Genetics Symp.* (Paestum, Italy) **1** 81-4
[5] Savaskan C, Szarejko I and Toker M C 1999 Callus production and plant regeneration from anther culture of some Turkish barley cultivars *T. J. Bot.* **23** 359-65
[6] Murashige T and Skoog F 1962 A revised medium for rapid growth and bioassays with tobacco tissue culture *Physiol. Plant* **15** 473-97
[7] Wilkinson M D, Castells-Brooke N and Shewry P R 2013 Diversity of sequences encoded by the Gsp-1 genes in wheat and other grass species *J. Cereal Sci.* **57** 1-9
[8] Minorsky P V 2002 The Wall becomes surmountable *Plant Physiol.* **128** 345-53.
[9] Kreuger M and van Holst G J 1996 Arabinogalactan proteins and plant differentiation *Plant Mol. Biol.* **30** 1077-86
[10] Freed R D 1991 MSTATC: Microcomputer statistical program. experimental design, data management and data analysis (Michigan, USA: Crop and Soil Department, Mich. Michigan State Univ.,)
[11] Bi R and Wang H 2008 Primary studies on tissue culture from mature embryos in diploid and tetraploid wheat *Front. Agric. China* **2** 262-5
[12] Wilkinson M D, *et al* 2017 The Gsp-1 genes encode the 1 wheat arabinogalactan peptide *J. Cereal Sci.* **74**, 155-64
[13] Jolly C J, Glenn G M and Rahman R 1996 GSP-1 genes are linked to the grain hardness locus (<I>Ha</I>) on wheat chromosome 5D *Proc. of the National Academy of Sciences* (USA) **93** 2408-13 (in Cataloque of Gene Symbols for Wheat. 10.IWGS, paestum, Italy 2003 Produced by McIntosh R A, Yamazaki Y, Devos K M, Rogers J and Appels R)
[14] Turner M, Muhai Y, Leroy B, Charef B, Appels R and Rahman S 1999 The <I>Ha</I> locus of wheat: Identification of a polymorphic region for tracing grain hardness in crosses *Genome* **42** 1242-8 (in Cataloque of Gene Symbols for Wheat. 10.IWGS, paestum, Italy 2003 Produced by McIntosh R A, Yamazaki Y, Devos K M, Rogers J and Appels R)
[15] Fincher G B, Stone B A and Clarke A E 1983 Arabinogalactan proteins: structure, biosynthesis and function *Ann. Rev. Plant Physi.* **34** 47-70
[16] Ma H and Zhao J 2010 Genome-wide identification, classification, and expression analysis of the arabinogalactan protein gene family in rice (Oryza sativa L.) *J Exp Bot* **61** 2647-68

Properties of different aged jicama (*Pachyrhizus Erozus*) plants

F Nursandi[1], M Machmudi, U Santoso and D Indratmi

Agrotechnolgy department, Muhammadiyah of Malang University, Indonesia

E-mail: fatnursandi@gmail.com

Abstract. Jicama crop potential is very large, the tuber is used as a fresh fruit, ice mix fruit, salad, and can be made into flour, starch and inulin. The nutritional content of yam tubers depends on the age of the harvest, while farmers harvest jicama tubers at the age varying between 4-6 months. The research objective is to analyze the content of proximate fresh tubers and three kinds of flour (flour, starch and starch dregs) by harvesting different age plants. The study was conducted in Malang at a height of 560 m above sea level. Planting was done using plastic mulch with a spacing of 80 cm x 20 cm. Research using complete Randomized block Design with one factor harvesting consisting of 16, 18, 20 and 22 weeks after planting. Jicama tubers were harvested and analyzed the proximate for moisture, ash, fat, protein and carbohydrates in the fresh tubers, flour, starch and jicama flour dregs. The results showed that the late harvest resulted in moisture content, ash content, fiber and fat increase while the protein and carbohydrate decreased. The content of carbohydrates in the flour, starch and starch dregs was almost the same at different harvest time. Flour contains carbo hydrates from 79.29 to 82.54%, starch from 86.35 to 89.40% starch and starchy dregs from 78.42 to 80.71%. The protein content of the flour is from 4.22 to 5.87%; while protein content of starch and protein content flour dregs is from 1.05 to 1.90% and 3.95 to 4.84%. Flour fiber content increased with increasing age of plants, while the fiber content of starch decreased but the dregs flour fiber content is almost the same

1. Introduction

The use of materials that can be renewed for a variety of human needs is likely to increase in future, because the ingredients are readily available, environmentally friendly and can be described [1]. These materials can be applied to the food industry, cosmetics and pharmaceuticals as a material emulsion, foam, texture and microencapsulation [1]. Yam crop potential is very large, the tuber is used as a fresh fruit, ice mix fruit, salad, and can be made of flour, starch and inulin. In the process of making starch produced pulp and can be made of flour. The results of study [2] states yam flour fiber or pulp can serve as a prebiotic and potential as a functional food component. The nutritional content yam tubers are affected by some factors, tuber phenotype [3] and the place of planting and harvesting [4,5]. Different harvestings produce different chemical contents, while farmers harvest at the age that varies between 4-6 months, depending on considerations of selling price in the market. Research relating to the yam flour has been done primarily and it related to the further process of making starch into material that is more specific to the desired function well for the food, pharmaceutical and cosmetics [1,3,6]. Jicama is not durable at room temperature or low temperature, so processing it into flour and starch will reduce damage and increase the sale value of yam tubers. This study aimed to analyze the chemical composition of proximate flour, starch and starch dregs yam tubers of different aged jicama plants.

2. Methods

The materials used in this study is the yam tubers with different harvest ages, plastic mulch, chemicals for proximate analysis. The study was conducted in Malang at a height of 560 m above sea level. Planting was done at a spacing of 80 cm x 20 cm using plastic mulch. Research was done using complete randomized block design with one factor of harvesting 16, 18, 20 and 22 weeks after planting. Each treatment was repeated three times.

Yam tubers were harvested and further treated into flour, starch and yam flour dregs. The process of making each one is as follows: first, yam flour. Yam flour-making process has several stages. Jicama was washed and peeled and then grated roughly using a coarse grater. Coarse grated yam tubers were dried for 3 x 24 hours in the blazing sun and subsequently dried in ovens at 80°C for 1 x 24 hours. Grated were blended and filtered using a fine sieve. Then it was analyzed into proximate flour. Second, yam starch. The process of making yam starch includes: washing yam, peeling it, and extracting by using juicer. Yam juice was then aged for 1 x 24 hours later, so the water was separated from the precipitate. Wet starch precipitate was dried in the sun for 2 x 24 hours. Precipitate was further roasted at a temperature of 80°C for 1 x 24 hours. The starch smoothed by using a blender and was filtered by using a fine sieve. Starch was then analyzed into proximate yam flour dregs. Yam dregs was obtained from the separation of the starch manufacturing process. Dregs were dried for 3 x 24 hours in the blazing sun and subsequently ovendried at a temperature of 80°C for 1 x 24 hours. Dregs are dried yam mashed with a blender and filtered by using a fine sieve. Yam flour dregs were analyzed proximately.

Proximate analysis included moisture content, ash content, fiber content, fat content, protein content and carbohydrate content. Measurement of moisture content was done by using the oven method [7], ash content using AOAC method in 2005 [8], crude fiber content by gravimetric method [9], the fat content by Soxhlet method [10], the protein content by the method Kjehdahl [7]. Data were analyzed by using ANOVA and followed by LSD test with α level of 5%.

3. Results and discussion

Proximate chemical composition of yam tubers affected by harvest time, the phenotype in one variety and planting. [3] reported a 5 phenotypes based on skin color yam tuber varieties Pachyrhizus tuberosus show proximate different chemical constituents of the fresh roots and flour. Tuber water content ranged from 70.96 to 87.45%, carbohydrate content from 3,58 to 21.27%, ash content from 1.58 to 2.49% and protein content from 4.35 to 7.433%. While [5] using *P. erosus* one phenotype harvested at different ages showed significant differences in ash content, protein and carbohydrate, while the water content and fiber content is no different. The ash content increased with increasing age of plants at harvest is 2.282% at 16 weeks be 5.752% at the age of 22 weeks. While the protein content of the opposite occurs, namely increasing age harvest decreased protein content is 3.731% at 16 weeks be 1.269% at the age of 22 weeks.

Table 1. Proximate chemical composition yam flour with different age plants.

Age (weeks)	Water (%)	Ash (%)	Fiber (%)	Fat (%)	Protein (%)	Carbohydrate (%)
16	9,89 ab	4,32 b	3,52 a	0,38 a	5,87 b	79,55 a
18	9,86 ab	2,89 a	6,07 b	0,29 a	5,86 b	81,10 b
20	8,29 a	3,50 a	6,16 b	0,45 a	5,24 b	82,54 b
22	11,86 b	4,34 b	9,27 c	0,29 a	4,22 a	79,29 a

Table 2. Proximate chemical composition yam starch with different age plants.

Age (weeks)	Water (%)	Ash (%)	Fiber (%)	Fat (%)	Protein (%)	Carbohydrate (%)
16	8,29 a	0,75 a	1,94 a	0,32 a	1,90 b	88,73 b
18	8,47 a	0,58 a	1,62 a	0,17 a	1,38 a	89,40 b
20	9,17 a	0,90 a	1,79 a	0,13 a	1,23 a	88,57 b
22	11,39 b	0,92 a	1,34 a	0,28 a	1,05 a	86,36 a

Besides consuming fresh yam tubers, these can be processed into flour and starch. In the process of making starch produced pulp juice and pulp results can be made of its flour. Yam harvest age proximate influence the chemical composition of flour, starch and dregs yam as shown in tables 1, 2 and 3. Age significantly affected the harvest moisture, ash, fiber, protein and carbohydrates while the fat had no effect. With increasing age of harvesting the fiber content of flour increased while protein content decreased. The same was reported by [4] but with different values for different cropping place [3]. The yam plant in Tanah Merah, Binjae, North Sumatra, at the age of 16 crops and 22 weeks obtained proximate chemical composition ie flour protein content of 9.83% and 8.45%, 20.29% and fiber content 22.93%. While in this study conducted in Malaysia with the same varieties are Gajah varieties obtained different results ie flour protein content of 5.87% and 5.24%, the fiber content of 3.52% and 6.16%. This shows that the planting of chestnuts can affect the proximate chemical composition of yam flour. Additionally varieties also influence the chemical composition of yam flour proximate, the P tuberosum varieties obtained protein content of 4.35% -7.43% [3].

The proximate chemical composition starch and flour yam showed difference. Ash, fat and protein fibers in the yam starch was lower than the yam flour as shown in tables 1 and 2. Harvesting time of yam significant effected the levels of protein and carbohydrates, moisture, ash, fiber and fat are not significantly affected. The protein content decreased with age, while the fiber in flour increase (Table 1). Harvesting time of yam significant effected the levels of protein and carbohydrates, while the moisture, ash, fiber and fat are not significantly affected. The protein content decreased with age, while carbohydrate in starch from tubers harvested from age 16-20 was not significantly different more than the carbohydrate in tubers harvested at 22 weeks a shown in table 2.

Table 3. Proximate chemical composition dregs yam residual starch with different age plants.

Age (weeks)	Water (%)	Ash (%)	Fiber (%)	Fat (%)	Protein (%)	Carbohydrate (%)
16	11,99	2,78	13,83	0,32	4,73	80,18
18	13,63	2,48	13,03	0,62	4,84	78,42
20	12,24	2,98	13,24	0,61	4,41	79,76
22	12,45	2,47	12,76	0,42	3,95	80,71

Dregs yam residual starch ranged between 17% - 22% of wet weight or 2.55% - 5.73% [11], and this value is quite large. Dregs quickly ferment and cause odor. Dregs can be made of flour and proximate chemical composition can be seen in table 3. Flour yam pulp fiber is high at 12.76% - 13.83%, this value is higher than the fiber content in flour and yam starch. In addition, dregs yam flour protein content is also quite high 33.95 to 4.84% as shown intable 3, the dregs yam flour can be used as animal feed mixes or mixed foodstuffs [2]. Consumption of fiber yam flour can improve the health of the colon and has potential as a functional food component. Furthermore, Purwandani L [2] stated fiber yam flour contains soluble dietary fiber 4.04%, 51.21% insoluble fiber, inulin 172 ppm. Consuming fiber yam flour reduces the population of E coli, increases the water content, total short chain fatty acid, lowers the pH and gives effect to pour in the stool. Based on the chemical composition of flour and starch proximate harvested at 18-20 weeks were better than those harvest date the age 22 of weeks. Latitude J A *et al* [4] reported a flour derived from yam tubers at harvest age of 3 months and 4 months yield methanolic extract, and more antioxidant activity than bengkoang optimum harvesting age 5 months. There is a slight difference in harvesting, allegedly caused by different places namely Malang and North Sumatra.

4. Conclusion

Flour fiber content increased with increasing age of harvest, while the fiber content of starch decreased but the dregs flour fiber content was almost the same. The proximate chemical composition of flour and starch harvested at 18-20 weeks was better than that harvested at the age of 22 weeks.

Acknowledgments

Thanks to Kemenristek Higher Education which has provided funding for the program melalaui in Leading Research Universities 2014

Reference

[1] Makkar R S, Cameotra S S and Banat I M 2011 Advances in utilization ofrenewable substrates for biosurfactant production AMB Express **1** 1-19
[2] Purwandani L 2011 Characteristics of the physical, chemical and physico-chemical fiber yam flour (Pachyrhizus erosus) as well as its potential as prebiotics, Gajah Mada University, Indonesia
[3] Ascheri J L R, Zamudio L H B, Carvalho C W P, Arevalo A M and Fontoura L M 2014 Extraction and characterization of starch fractions of Five phenotypes Pachyrhizus tuberosus (Lam.) Spreng Food and Nutrition Science **5** 1875-85
[4] Latitude J A, Rusmarilin H and Lubis L M 2014 Yam tuber extract antioxidant activity at different harvest method of DPPH (2,2-diphenyl-1-picrylhydrazyl) J. Rekayasa Food and Pert. **2(1)** 47-56
[5] Nursandi F, Machmudi M and Santoso U 2015 Studies yam harvest (pachyrhizus erozus) against flour and starch production by double cropping system *Proc. of the Annual Forum of National Development of Science and Innovation* 2015 **2088-2645** 309-13
[6] Carrera M'arquez E G, Aguirre-Mandujano E, Rodr'ıguez-Huezo M E and Alvarez-Ramirez J 2014 Stabilizingoil-in water emulsion with yam bean (pachyrhizus erosus L. URBAN) solids Revista Mexicana de Ingeniería Quimica **13(2)** 447-56
[7] Maharani E T and Yusrin Y 2010 Protein levels kista atemia Rainfall sold city farmers rembang with storage temperature variations *Proc. of the National Seminar, Muhammadiyah Semarang University, Indonesia*
[8] Amelia R M, Nina D, Trisno A, Julyanty S W, Rafika N F and Yuni H A 2014 Determination of levels of Abu (AOAC 2005) (Bogor, Indonesia: Department of Community Nutrition, Faculty of Human Ecology)
[9] Setyowati R, Sarbini D and Fortune S 2008 Effect of addition bekatul against crude fiber content, appearance personality and power on making tempe thank soybean (glycine max (L) meriil) Research Journal of Science & Technology **9(1)** 52-61
[10] Prasetyo R P, Santosa S S and Iriyanti N 2013 Use of functional level against fat content feed and broiler chicken meat protein Scientific Journal Ranch **1(1)** 289-98
[11] Machmudi F and Indratmi Nursandi D 2015 Optimization of extraction and characterization of inulin bengkuang results cultivation with plastic mulch (University Muhammadiyah of Malang)

Review on sugar beet salt stress studies in Iran

S Khayamim[1, 4], H Noshad[1], M R Jahadakbar[2] and K Fotuhi[3]

[1] Sugar Beet Seed Institute, Agricultural Research Education and Extension Organization (AREEO), Karaj, Iran
[2] Sugar Beet Research Department, Isfahan Agricultural and Natural Resources Research and Education Center, Agricultural Research Education and Extension Organization (AREEO), Isfahan, Iran
[3] Sugar Beet Research Department ,West Azarbaijan Agricultural and Natural Resources Research and Education Center, Agricultural Research Education and Extension Organization (AREEO), Urumie, Iran

E-mail: samar.khayam@gmail.com

Abstract. Increase of saline lands in most regions of the world and Iran, limit of production increase based on land enhancement and also threat of saline water and soils for crop production make related researches and production of salt tolerant variety to be more serious. There have been many researches about salt stress in Sugar Beet Seed Institute of Iran (SBSI) during several years. Accordingly, the new screening methods for stress tolerance to be continued based on these researches. Previous researches in SBSI were reviewed and results concluded to this study which is presented in this article in three categories including: Agronomy, Breeding and Biotechnology. In agronomy researches, suitable planting medium, EC, growth stage and traits for salinity tolerance screening were determined and agronomic technique such as planting date, planting method and suitable nutrition for sugar beet under salt stress were introduced. Sand was salinizied by saline treatments two times more than Perlit so large sized Perlit is suitable medium for saline studies. Sugar beet genotypes screening for salt tolerance and should be conducted at EC=20 in laboratory and EC= 16 dS/M in greenhouse. Although sugar beet seed germination has been known as more susceptible stage to salinity, it seems establishment is more susceptible than germination in which salinity will cause 70-80% decrease in plant establishment. Measurements of leaves Na, K and total carbohydrate at establishment stage would be useful for faster screening of genotypes, based on high and significant correlation of these traits at establishment with yield at harvest time. In breeding section, SBSI genotypes with drought tolerance background would be useful for salinity stress studies and finally there is a need for more research in the field of biotechnology in Iran.

1. Introduction

Salt stress is one of the most important abiotic stresses, which decreases crop yields and limits the land use. Three hectares of agricultural land are being destroyed in each minute by salinity [1]. 53 million hectares of salt affected grounds in Asia and Middle East (Iran, Syria, Egypt and Tunisia) are caused by man-made activities in irrigated fields [1]. In Iran, salinity is one of the most important factors which limit the agriculture. Mean of yield loss is estimated more than 50% in saline lands of Iran [1] which causes more than one milliard dollars financial loss annually [2]. Saline lands are increased to 33 million hectares which is 20% of total irrigated lands in Iran [3]. Salt stress studies were very important in the most sugar beet planting regions of Iran based on sugar beet experts and farmers opinions [4] Producing sugar beet salt tolerant varieties, improving planting methods, irrigation and fertilizer use are the most Sugar Beet Seed Institute (SBSI) strategies for optimizing the production in saline lands [5].

A lot of researches have been conducted in this institute, during several years which are categorized in Agronomy, Plant breeding and Biotechnology. In this article we reviewed these researches and summarized the most important factors.

2. Agronomy researches

2.1. Base agronomic research

There are different researches on medium of saline studies. Although sand is cheap and accessible medium [6], Seedling growth on Perlit, especially large size, was stronger and more suitable than sand because sand was salinized two times more than Perlit in salt treatment [7, 8].

Although seed germination in EC=16 was the same as EC=20 dS/m under laboratory conditions, 50 % seedling losses was observed at EC=16 dS/m in the greenhouse condition [7].

There are reports that salt stress didn't affect sugar beet till 35 [9] and 50 days growth stages [10]. Also germination of Sugar beet seeds decreased to 35 % and dead seedlings increased to 80% under salt stress (EC=16 dS/m) in greenhouse condition. So salinity decreased seed germination and by increasing dead seedlings, caused sugar beet establishment and final yield to be reduced. It is concluded that establishment stage in comparison to germination is more susceptible to salinity [11].

Breeding for salinity tolerance requires the study of different physiological mechanisms based on complexity of this stress [12]. Salt as the same as drought stress, has similar physiological acclimation mechanism. Thus development in drought resistance would be useful for plant adaptation to salinity, but it is important to pay attention that a salt tolerant plant should regulate toxic levels of NaCl by its uptake and compartmentalization [13]. The first important characteristic of seed quality is germination which shows significant correlations in laboratory with field tests [14]. Sugar beet seed germination and establishment are more affected by genetic effects than germination rate [15], so they could be used for genotype selection [16]. Root length and abnormal seedling in addition to the seed germination, are suitable traits for screening sugar beet salt tolerant genotypes at the first growth stages [11]. Crop production is the main trait for genotype selection in abiotic stress such as salinity but its inheritance is complicated and extremely influenced by environment [17]. There are various mechanisms in plants for adjusting to osmotic and ionic stresses. Accumulation of important solutes such as poly amins (glycine betaine and proline), ions (K^+) and carbohydrates are the main ways in crops [18, 19]. Leaf potassium, sodium, and their ratio, soluble carbohydrate, were more significant and related characteristics with yield under saline condition during whole growth stages [20, 21].

2.2. Practical Agronomic research

Practical agronomic methods such as suitable planting pattern, plant density, fertilization and crop protection are the most useful ways to increase yields under salinity stress. It is reported that yield of sugar beet transferred plants under salinity condition was more than direct planting [22]. Earlier and two-rows plantings were suitable methods in saline condition because of better irrigation and lack of accumulated salt on top of rows [22]. 25% more use of nitrogen fertilizer in saline land in comparison with normal condition [23] and also use of urea in comparison with ammonium sulfate were recommended for sugar beet growth in saline land of Iran [24]. As environmental conditions are very important, study on applied agronomic factors which affecting salinity stress would still be necessary for future.

3. Breeding researches

One of the most important strategies that will increase crop yields is breeding of new varieties. The complexity of improvement salt tolerant variety is because of multi-genic characteristic of salinity tolerance [25] which is genetically complicated character [26]. Also traits such as epistasis, genotype - environment interaction and great environmental effects cause the complexity of breeding for salt tolerant variety. So the most important way for breeding is utilization of genetic mechanism [27]. Salt tolerance indicates dominance, heterosis and also additive effects [12]. Sugar beet genotypic variation in drought [28, 29] and salinity tolerance [22, 30] are high that would be used for improving salt tolerant varieties. Also applying *Beta vulgaris ssp. Maritime* as a wild relative of sugar beet will be useful for making its gene pool rich. Both sugar beet female and male parental plants must be tolerant but improving of tolerant female parents is long lasting, so the activity for development of tolerant male parent is more than the female one [22]. Fortunately, there are chances for improving sugar beet tolerant varieties because of variation in germplasm especially it is observed that sugar beet drought tolerant genotypes were also tolerant to salinity stress [21, 31]. This is because of same physiological mechanism between these two stresses [13], concluding germplasm with drought tolerance back ground could be used for salt tolerance improvement.

4. Biotechnology researches

Conventional breeding programs are time lasting and somehow expensive because of multi-genic tolerance of abiotic stress so, applying advanced molecular techniques show promising results in understanding molecular background of abiotic tolerance and enhancement of this tolerance in crops. Results show the importance of molecular, physiological and metabolic studies in future to assist crop improvement with an inherent tolerance capacity [12]. Some molecular markers, such as SSR markers or microsatellite are suitable methods for describing genetic variation at DNA level [32]. 23 SSR markers were used to analyze genetic diversity in 168 sugar beet lines and many SSR marker polymorphisms was seen in these breeding lines [31]. If this marker applied repeatedly, it will establish marker-assisted selection for tolerance of salinity in this crop and may be used to estimate the potential heterosis from the distance between parents in hybrid breeding programs [12].

Proteomics is a method which is used in assessment of plants under stress condition. Several proteins with various roles and candidate genes which respond to salinity are identified by proteome and transcriptome analysis in monosomic addition line of sugar beet which would be useful for more studies of salt tolerance functions in future [33]. Some candidate proteins in sugar beet were observed in related to drought [34] and salinity tolerance [35] with proteomics approach. Also there is some quantitative trait loci (QTLs), related with salt tolerance in crops but there isn't any reports on sugar beet [12].

Molecular markers clarify quantitative genetic and use for facilitating selection effectively [36]. As told before, genetic of salt tolerance is complicated because of multiple genes and loci distributed at different locations through the genome of crops. The mixture of genes and alleles and their interactions would have great importance in comparison to gene expression [37]. After time, breeders should improve salt tolerance in plants through various biological methods [12].

Acknowledgments

We thank Ms. Fatemeh Baba for her comments and English editing on manuscript. This research was supported by Sugar Beet Seed Institute.

References

[1] FAO AGL 2000 http://www.fao.org/ag/agl/agll/spush
[2] Qureshi A S, Qadir M, Heidari N, Tural H and Javadi A 2007 A review of management strategies for salt prone land and water resources in Iran *Working Paper 125* (Iran: International Water Management Institute)
[3] Anonymous 2012 Iranian static centerhttp://salnameh.sci.org.ir
[4] Taleghani D, Sadeghzadeh Hemayati S and Mesbah M 2007 *Sugar Beet Seed Institute* **520** (In Persian)
[5] Sadeghzadeh Hemayati S and Khayamim S 2008 *Sugar Beet Seed Institute* **88614** (in Persian,

abstract in English)

[6] Beatty K D and Ehlig C F 1973 *J. A. S. S. B. T.* **17(4)** 295-9

[7] Khayamim S, Tavakol Afshari R, Sadeghian Motahar S Y V and Pustini K 2011 *Iranian J. Crop Sci.* **13(1)** 1-17 (in Persian, abstract in English)

[8] Khayamim S and Noshad H 2012 *Proc. of 2nd National Congress on Hydroponic and Green House Products* (Mahalat, Iran) pp 342-3 (in Persian, abstract in English)

[9] Niazi B H, Athar M and Rozema J 2004 *Bulgarian J. Plant Phy.* **30** 78-88

[10] Delfine S, Alvino A ,ConcettaVilani M and Loreto F 1999*Plant Phy.* **119** 1101-6

[11] Khayamim S, Tavakol Afshari R, Sadeghian Motahar S Y, Pustini K, Rouzbeh F and Abbasi Z 2014 *JAST* **16(2)** 779-90

[12] Rajabi A, Khayamim S, Abbasi Z and Ober E 2014 *Improvement of Crops in the Era of Climatic Changes* ed P Ahmad, M R Wani, M M Azooz and L S Phan Tran (London Springer) pp 21-150

[13] Munns R 2002 *Plant, Cell and Env.* **25** 239-50

[14] Sadeghian S Y and Yavari N 2004 *J. Agr. Crop Sci.***190** 138-44

[15] Sadeghian S Y and Khodaii A 1998 *Euphytica* **103** 259-63

[16] Mohammadian R 1995 Effects of seed exhausted in germination, seedling and establishment of seven sugar beet lines under salt stress (Iran: Tabriz University) (in Persian with English abstract)

[17] Ashraf M 2004 *Flora* **199** 361-76

[18] Munns R 2005 *New Phytol* **167(3)** 645-63

[19] Hasegawa P M, Bressan R A, Zhu J K and Bohnert H J 2000 *Ann Rev Plant Biol* **51(1)** 463-99

[20] Khayamim S 2010 Study of physiological parameters of sugar beet lines in salt stress and normal conditions (Iran: University of Tehran) (in Persian, abstract in English)

[21] Khayamim S 2014 *Sugar Beet Seed Institute* **9346693** (in Persian, abstract in English)

[22] Rajabi A 2010 *Sugar Beet Seed Institute* **90.123** (in Persian)

[23] Jahadakbar M R 2005 *Sugar Beet Seed Institute* **841396** (in Persian, abstract in English).

[24] Shahabifar J 2008 *Soil and Water Research Instiute Iran* (in Persian, abstract in English)

[25] Munns R 2011 *Advances in Botanical Research* ed T Ismail (Massachusetts: Academic Press) pp 1-32

[26] Shannon M C 1985 *Plant Soil* **89** 227-241

[27] Joshi A K 2011 *Handbook of Plant and Crop Stress* ed M Pessarakli (Florida: CRC Press) pp 803-50

[28] Rajabi A and Ober E S 2012 *Crop Production for Agricultural Improvement* ed M Ashraf, M Ozturk, M S Aqil Ahmad and A Aksoy (London: Springer) p 796

[29] Rajabi A, Vahidi H, Haj Seyed Hadi M R and Fathollah Taleghani D 2013 *Int J Agric Crop Sci* **5-7** 761-8

[30] Ober E S and Rajabi A 2010 *Sugar Tech***12(3-4)** 294-8

[31] Abbasi Z, Arzani A, Majidi M, Nachtigall M and Frese L 2012 *Joint Meeting of ECPGR and WBN* (Lille France) p 21

[32] Kalia R K, Rai M, Kalia K, Singh S and Dhawan R A K 2011 *Euphytica* **177** 309-34

[33] Yang L, Ma C, Wang L, Chen L and Li H 2012 *J Plant Phy.* **169(9)** 839-50

[34] Hajheidari M, Abdollahian-Noghabi M, Askari H, Heidari M, Sadeghian S Y, Ober E S and Salekdeh G H 2005 *Proteomics* **5(4)** 950-60

[35] Khayamim S 2016 *Proc. of 75th IIRB Congress* (Brucelss Belgium) p 104

[36] McGrath J M 2011 *Sugar Tech* Doi:10.1007/s12355-010-0051-3

[37] Stevanato P, Trebbi D and Saccomani M 2010 *Euphytica* **173** 289-98

Unusual varieties and duplication of Rig-I like receptors encoded in the marine mollusk, *Crassostrea gigas*

Z H Tian and C Z Jiao[1]

Life Science College, Shaoguan University, Shaoguan, 512005, China

E-mail: jiaocz@gmail.com

Abstract. RIG-I like receptors (RLRs) play key roles in sensing non-self nucleic acids in cytoplasm and trigger antiviral innate immune response in vertebrates and human body. Here we carried out *in silico* analysis to identify and investigate the putative RLRs encoded in the genome of marine mollusk, *Crassostrea gigas* (cgRLRs), an invertebrate species. We found the unusual duplication and varieties on domain architecture of putative cgRLRs encoded in the genome of *C. gigas*. Three putative cgRLRs (accessions numbers are EKC24603, EKC31344.1 and EKC38304.1 on GenBank), have the similar domain architecture with that of human RIG-I or MDA5, and one protein (EKC34573.1) with that of human LGP2; The fifth putative cgRLRs (EKC38303.1) is somewhat similar with human RIG-I/MDA5 except that it has only one caspase activation and recruitment domain (CARD) in its N-terminal. Other nine proteins were identified to be partialy similar with RLRs while with the incomplete sequences, which maybe reflect the events of partial duplication of cgRLRs genes occurred in the oyster genome.

1. Introduction

Animals have developed sophisticated mechanisms to detect and combat intrusions of pathogenic microbes in their long evolutionary history. Innate immune receptors in certain cells are responsible for recognizing pathogen-associated molecular patterns (PAMPs), which are structures or chemical patterns that are extensively conserved in bacterial cell wall components, microbial nucleic acids, and certain highly conserved proteins. These innate immune receptors are classically termed patter-recognition receptors (PRRs). Upon their recognition of PAMPs, PRRs trigger downstream signaling that rapidly activate and amplify potent inflammatory antimicrobial responses, mainly through the upregulation of the gene expression and activation of antimicrobial molecules and many cytokines, including NF-κB, type I interferons, or caspases 1 [1].

There are at least three classes of PRRs being identified as sensors of viral nucleic acids: Toll-like receptors (TLRs) which sense viral RNA or CpG-containing DNA in endolysosomes; Viral DNA sensors of structurally diverse group, such as cGAS, IFI16, and DAI; RIG-I-like receptors (RLRs), which sense viral RNA in the cytoplasm of most cell types [2]. In human body, three RLRs members including RIG-I, MDA5, and LGP2 have been identified and studied up to now [2]. RLRs have conserved domain structure. The RIG-I and MDA5 consist of two tandem N-terminal caspase recruitment domains (CARD1 and CARD2) with death domain folds, a DExD/H-box helicase (consisting of two RecA-like helicase domains, Hel1 and Hel2, and an insert domain, Hel2i), and a C-terminal domain (CTD); The third RLR, referred to LGP2, has similar DExD/H-box helicase and CTD conserved domains than that of RIG-I and MDA5, lacks CARD domains in its N-terminal [3,4].

Both RIG-I and MDA5 bind dsRNA through interactions were mediated by the CTD and the DExD/H-box helicase domain, while the different target recognition modes at the CTD allow the two proteins to sense non-self RNAs of different structural patterns [5,6]. The RIG-I senses RNAs bearing a triphosphate (ppp) moiety in conjunction with a blunt-ended, base-paired region at the 5'-end, and it also senses RNAs bearing 5'-diphosphates (5'pp) as shown in a recent study [7]. The CTD of RIG-I caps the duplex terminus through stacking interactions with terminal base pairs [8]. The RIG-I will be active in ATP hydrolysis and downstream signaling after binding of its target RNA [5]. Different from RIG-I, the target of MDA5 is long and accessible RNA duplexes [9]. The CTD of one MDA5 forms a cooperative binding interface with the insertion domain of a neighbor and leads to the oligomerization of MDA5, resulting in formation of long filaments along RNA duplexes [8,10]. The filaments of MDA5 and dsRNA may cause clustering of the CARDs of MDA5, making them active in taking part in immune signaling. LGP2 is present at low levels in the uninfected cells but accumulates in response to virus infection or antiviral mediators, including poly (I-C) and IFNs. Due to absence of CARD or any other signaling interaction domain, the function of LGP2 is largely unknown, while the accumulating evidence shows that it is intimately associated with MDA5 as accomplice and antagonists of antiviral signal transduction [3].

Our understandings of the RLRs are mostly obtained from research on mammals and human. For invertebrates, relatively little is known about their RLRs, if they exist, and their function. The Pacific oyster *Crassostrea gigas* (Thunberg, 1793) is a marine bivalve of the phylum Mollusca, which comprises the largest number of described marine animal species. Many molluscs are important fishery and aquaculture species, and are also models of research in areas, such as neurobiology, biomineralization, ocean acidification and adaptation to coastal environments under climate change [11,12]. Using the genome sequence of *C. gigas* released in 2012 [13], we carried out *in silico* analysis to identify and investigate the putative RLRs coded for in the oyster genome using bioinformatics methods. We found multiple members of putative RLRs, with a variety of domain architectures, encoded within the *C. gigas* genome, which is much more abundant than that in genomes of vertebrates and humans.

2. Methods

2.1. Database mining

To identify putative RLRs (RIG-I, MDA5 or LGP2) encoded in the genome of *C. gigas* (cgRLRs), we employed BLASTP and tBLASTn [14] searches of the human proteins RIG-I, MDA5 and LGP2 from non-redundant protein sequences (nr) and whole-genome shotgun contigs (wgs) database on GenBank. The hits were manually checked based on scores and annotation in GenBank and selected as candidate cgRLRs or genes encoding them to be analyzed further.

2.2. Sequence analysis

The conserved domains of candidate cgRLRs were predicted by searching against the database CDD [15] using a specialized blast (http://www.ncbi.nlm.nih.gov/Structure/cdd/wrpsb.cgi). The full-length amino acid sequences and the conserved functional domains were used in phylogenetic tree analysis. Multiple protein sequence alignments were performed using the MUSCLE [16] programme in MEGA (6.0) [17], which were manually curated if necessary. Phylogenetic trees were constructed using the neighbour-joining method within the MEGA (6.0) package. Data were analyzed using poisson correction, and gaps were removed by pairwise deletion. The topological stability of the neighbour-joining trees was evaluated by 1000 bootstrap replications.

2.3. Genomic structure analysis

All the genomic data used in this analysis were based on Entrez genome and Entrez nucleotide databases. Protein sequences were taken from Entrez protein databases on National Center for Biotechnology Information (NCBI) (https://www.ncbi.nlm.nih.gov/).

IOP Conf. Series: Earth and Environmental Science **77** (2017) 012005 doi:10.1088/1755-1315/77/1/012005

3. Result

3.1. Sequence identification

With the known human RIG-I protein sequence (NP_055129.2) as query, we performed extensive BLAST analysis of the genome, proteins, and EST databases of the pacific oyster in GenBank. Thirteen proteins annotated as RIG-I family members were identified (table 1).

Table 1. Proteins analyzed in this paper. Blast result is of the human RIG-I protein sequence (NP_055129.2) as query to search the *C. gigas* protein database in GenBank.

GenBank Acc.	Annotation	Max score	Total score	Query cover	E-value	Identity	Length (aa)	DBSOURCE
EKC24603.1	DDX58	360	360	72%	1.00E-106	35%	1198	JH817706.1
EKC38303.1	DDX58	348	348	74%	3.00E-104	34%	948	JH819181.1
EKC31344.1	DDX58	286	286	96%	1.00E-81	28%	1018	JH816862.1
EKC34573.1	DDX58	264	264	75%	2.00E-74	30%	902	JH822697.1
EKC38304.1	IFIH1	238	411	73%	6.00E-65	35%	1260	JH819181.1
EKC37415.1	DHX58	169	169	59%	2.00E-44	27%	538	JH816918.1
EKC34571.1	DDX58	164	164	47%	4.00E-43	30%	440	JH822697.1
EKC29027.1	DDX58	137	137	38%	4.00E-34	30%	459	JH818839.1
EKC24604.1	IFIH1	134	258	46%	3.00E-32	35%	307	JH817706.1
EKC21486.1	Fanconi anemia group M protein	95.1	95.1	56%	1.00E-19	24%	2327	JH817053.1
EKC23131.1	DDX58	69.3	69.3	7%	4.00E-13	46%	158	JH818146.1
EKC29026.1	IFIH1	70.5	70.5	17%	6.00E-13	33%	313	JH818839.1
EKC26346.1	Endoribonuclease Dicer	58.2	103	31%	2.00E-08	31%	1759	JH817198.1
EKC33722.1	Dicer-like protein 2 *	56.2	99	18%	6.00E-08	45%	520	JH819141.1
EKC34572.1	DHX58	49.7	49.7	19%	3.00E-06	25%	270	JH822697.1
EKC31345.1	IFIH1	38.1	38.1	7%	0.018	30%	365	JH816862.1

DDX58, Putative ATP-dependent RNA helicase DDX58, also known as RIG-I
DHX58, Putative ATP-dependent RNA helicase DHX58, also known as LGP2
IFIH1, Interferon-induced helicase C domain-containing protein 1, also known as MDA5
* Also annotated as a putative cgRLRs in this study.

Three related oyster proteins annotated as Fanconi anemia group M protein, Endoribonuclease Dicer, and Dicer-like protein 2 were also included in this study as control due to their high score of BLAST results. We noted that the score of target EKC31345.1 is very low among the listed targets while it is annotated as MDA5. With human MDA5 (NP_071451.2) and LGP2 (NP_077024.2) as query to search, we found the similar targets as above and no other protein was identified annotated as member of RIG-I family. The genomic sequences corresponding with each RIG-I family proteins were also analyzed and will be discussed later in this paper.

3.2. Domain analysis

The conserved domains of putative cgRLRs and human RLRs were predicted by searching in the database of CDD v3.11 with an expect value threshold of 0.01 (figure 1).

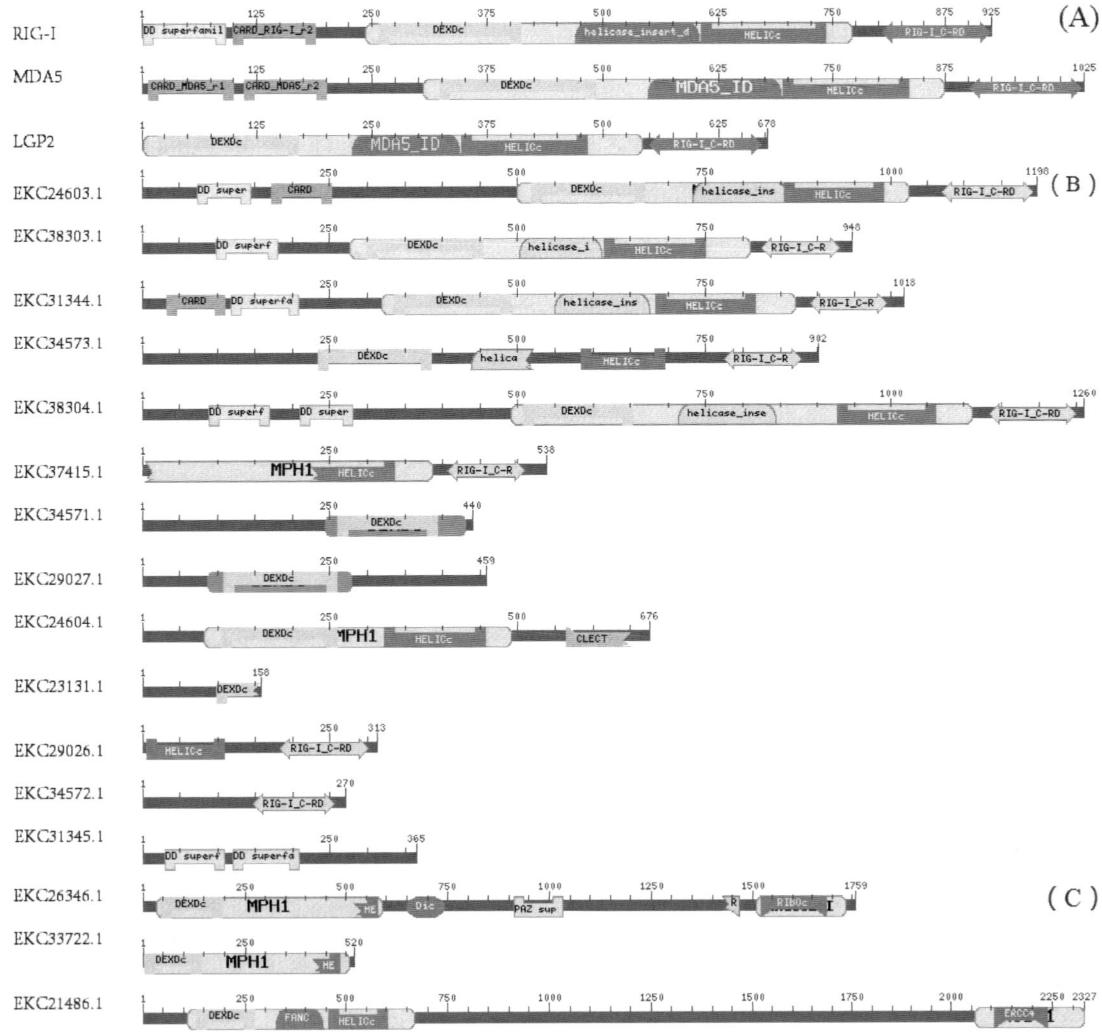

Figure 1. Domain architecture of putative cgRLRs and other closely related proteins. (A) human RLRs; (B) putative cgRLRs; (C) Endoribonuclease Dicer (EKC25346.1), Dicer-like protein 2 (EKC33722.1) and Fanconi anemia group M protein (EKC21486.1) of *C. gigas*.

Three putative oyster cgRLRs, which include EKC24603.1, EKC31344.1 and EKC38304.1 have the same architecture as human RIG-I or MDA5, which share all the core protein domains or conserved regions including two Caspase activation and recruitment domains (CARD or DD superfamily), the DEAD-like helicases superfamily (DEXDc or Hel1), helicase insert domain (Hel2i), Helicase superfamily c-terminal domain (HELICc or Hel2), and C-terminal domain of RIG-I (RIG-I_C-RD or CTD). Two other proteins, EKC38303.1 and EKC34573.1, have the same architecture as the three proteins above except that EKC38303.1 has only one CARD predicted, and no CARD predicted on EKC34573.1. The protein EKC34573.1 shares the same domains with human LGP2, while it contains a sequence of about 230 amino acids on the N terminal upstream of the DEXDc/Hel1 domain, which was not predicted as any conserved domains in this study. Three other proteins, EKC37415.1, EKC29026.1 and EKC34572.1 seem to be only the C terminal of RLRs. The proteins EKC37415.1 and EKC29026.1 contain the HELICc/Hel2 and RIG-I_C-RD/CTD domains. The protein EKC34572.1 contains the RIG-I_C-RD/CTD domain only. The proteins of EKC34571.1, EKC29027 and EKC23131.1 were all predicted to contain DEXDc/Hel1 conserved region only. The

IOP Conf. Series: Earth and Environmental Science **77** (2017) 012005 doi:10.1088/1755-1315/77/1/012005

protein of EKC31345.1 was predicted to contain two CARD domains. It was surprising that the protein EKC24604.1, which was annotated as Interferon-induced helicase C domain-containing protein 1(MDA5), contains a DEXDc/Hell domain, a HELICc/Hel2 domain, and a C-type lectin (CTL)/C-type lectin-like (CTLD) domain (CLECT). The CLECT domain is not contained in the other known RIG-I family proteins. The oyster Endoribonuclease Dicer (EKC26346.1) and Fanconi anemia group M protein (EKC21486.1) have a different architecture each with the RLRs, while they all share the similar domains of DEXDc and HELICc. The protein EKC33722.1 which was annotated as Dicer-like protein 2 on GenBank, contains DEXDc and HELICc domains. We performed the blastp searches with EKC33722.1 as query in human reference proteins database and found that the best hit was human MDA5 (NP_071451.2, with the E- value of 1e-18), so it was also considered as a putative cgRLRs. This was also supported by the following phylogenic analysis.

3.3. Phylogenic analysis

To investigate the relationships of the oyster RIG-I family proteins with those of other vertebrate organisms, the phylogenic tree was constructed with selected available RLRs of vertebrate organisms and the putative cgRLRs (figure 2).

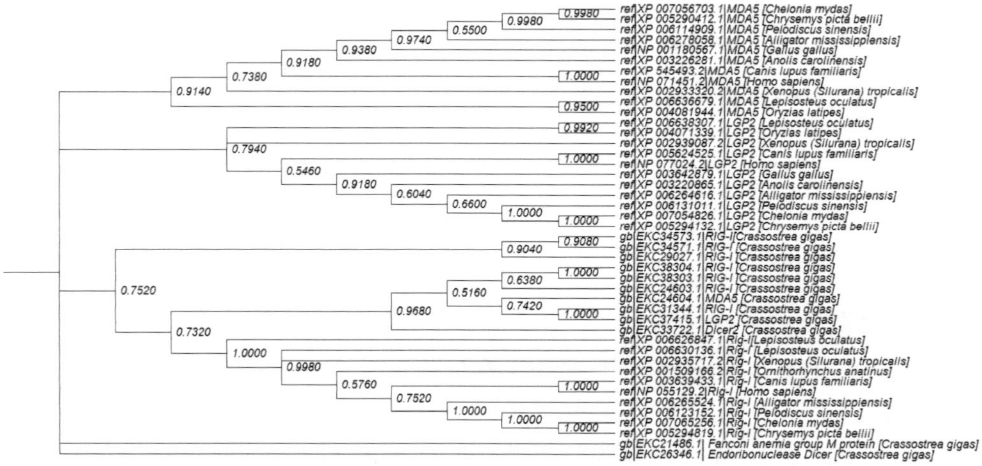

Figure 2. Phylogenetic tree analysis of RLRs.

Three related proteins of *C. gigas,* Dicer-like protein 2, Fanconi anemia group M protein, and Endoribonuclease Dicer were also included in this analysis. All sequences were downloaded from GenBank and aligned with MUSCLE. The trees were constructed using the neighbour-joining method within the Mega 6.06 programme. The bootstrap values of the branches were obtained by testing the tree 1000 times and values over 50% are indicated on the tree.

The phylogenic tree shows that RLRs and cgRLRs were clustered into three groups. Ten of the oyster proteins were clustered into the same group with the vertebrate RIG-I proteins regardless their annotation on GenBank as RIG-I, MDA5, LGP2 or Dicer2. Three other oyster proteins, EKC23131.1, EKC29026.1, EKC34572.1 and EKC31345.1 (figure 1) are not included in this tree due to their sequences being too short. To investigate the relationships of CARD domains in the putative cgRLRs, phylogenetic trees of CARD domain sequences were constructed (figure 3). It is apparent that the CARD of EKC38303.1 and the CARD2 of EKC38304.1, the CARD1/2 of EKC31344.1 and EKC31345.1 are the closest related among the CARDs analyzed (figure 3).

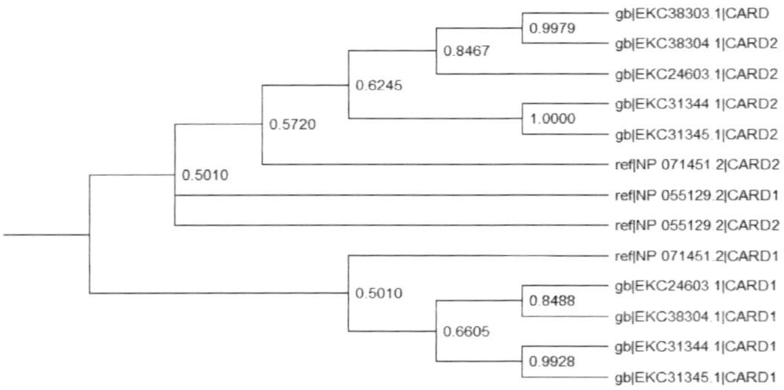

Figure 3. Phylogenetic tree analysis of CARD domains of RLRs.

The sequence of CARD domains of RLRs were predicted by searching in the database of CDD v3.11 with an expect value threshold of 0.01. Sequences of CARDs were aligned with MUSCLE and manually curated. Tree was constructed using the neighbour-joining method within the Mega6.06 programme. The bootstrap values of the branches were obtained by testing the tree 1000 times and values over 50% are indicated on the tree.

3.4. Gene Structure Comparison

Here we investigated the genomic composition of genes encoding cgRLRs, by focusing on the number and location of introns related to the CDS regions that encode different domains (figure 4).

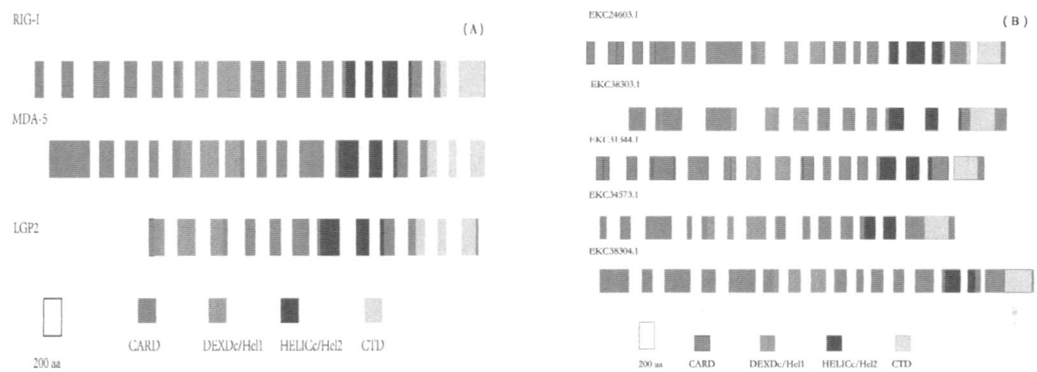

Figure 4. Genomic structure of RLRs in human (A) and *C. gigas* (B).

Major domains are marked with rectangles of different color. The transcript coding region is drawn to scale and introns (blank) positions are illustrated but their size is not drawn to scale.

The gene sequence encoding the CARD1 domain of human RIG-I is split by two introns, while that of human MDA5 is located in one exon. The CARD1 domains of putative cgRLRs are split by one (EKC24603.1, EKC38304.1) or two introns (EKC31344.1). The CARD2 domains of human RIG-I and MDA5 are both split by one intron, while those CARD2 domains of putative cgRLRs (EKC24603.1, EKC38304.1 and EKC31344.1) are all located in one exon. The protein EKC38303.1 contains only one CARD, which like CARD2 domains of other cgRLRs, located in one exon.

The DEXDc/Hel1 domain of human RIG-I is split by three introns, and those of human MDA5 and LGP2 are both split by two introns. For oyster, only the DEXDc/Hel1 domain of EKC24603.1 is split by two introns and this domain of other proteins (EKC38303.1, EKC31344.1, EKC34573.1 and EKC38304.1) are split by three introns. The HELICc/Hel2 domain of human RIG-I are split by 3 introns while that of MDA5 or LGP2 is split by 2 introns. The HELICc/Hel2 domain of cgRLRs is split by 2 introns (EKC24603.1, EKC38303.1, EKC31344.1, and EKC34573.1) or one intron (EKC38304.1). The RIG-I_C-RD/CTD of human RIG-I is split by one intron while that of MDA5 or LGP2 is split by two introns. The RIG-I_C-RD/CTD of oyster protein EKC24603.1 is split by one intron, other putative cgRLRs (EKC38303.1, EKC31344.1, EKC34573.1 and EKC38304.1) have no intron in the genome sequence encoding their RIG-I_C-RD/CTD domains.

Even though the genome sequence of Oyster is not assembled into contigs corresponding to each chromosome yet, we could still see genes which are vicinity in the genome using the information of their sequence source (table 1). The cgRLRs whose genes are clustered into together in genome are EKC24603.1 & EKC24604.1, which are both located in scaffold1785 (locus JH817706); EKC38303.1 & EKC38304.1, which are both located in scaffold348 (locus JH819181); EKC34572.1 & EKC34573.1, which are both located in scaffold609 (Locus JH822697); EKC31344.1 & EKC31345.1, which are both located in scaffold983 (Locus JH816862); EKC29026.1 & EKC29027.1, which are both located in scaffold1228 (Locus JH818839); EKC34571.1 & EKC34572.1 & EKC34573.1, which are all located in scaffold609 (Locus JH822697).

4. Discussion

The mechanisms of non-self-nucleic acids being recognized by PPRs and triggering signals of IFN production and antiviral responses are evolutionarily conserved from teleost fish to mammals.[18,19]. However, it is generally accepted that IFN-mediated antiviral immunity is absent in invertebrates, as the genes coding IFN homologous and it's major effectors, interferon-stimulated genes (ISGs), are absent in several fully sequenced invertebrate genomes [20, 21]. This view on invertebrate antiviral immunity maybe updated as more studies on antiviral mechanisms in Pacific oyster or other bivalve are available. Earlier studies showed that a STAT-like pathway in *Mytilus galloprovinciali*s haemocytes is activated following the treatment with interferon-omega [22]. Many genes encoding homologous to mammalian ISGs, including PKR, IRF, Mpeg1, and IFI44, have been reported to be involved in oyster antiviral response. An "IFN-like pathway" may exist in mollusks [21].

Our research shows that there are at least 14 putative RLRs encoded in the *C. gigas* genome. Seven of them (EKC37415.1, EKC34571.1, EKC29027.1, EKC2313.1, EKC29026.1, EKC34572.1, EKC31345.1, and EKC33722.1) seem to be partial segments of the complete RLRs and contain part of the conserved domains of mammalian (human) RIG-I, MDA5 or LGP2; Three of the putative cgRLRs (EKC24603.1, EKC31344.1, EKC38304.1) have a similar domain architecture to that of mammalian RIG-I or MDA5; The protein EKC34573.1 has no CARD domain and its domain composition and arrangement is similar to that of mammalian (human) LGP2. The protein EKC38303.1 is unique as it contains only one CARD in its N-terminal, while mammalian RIG-I and MDA5 both have two N-terminal CARDs. This type of domains composition is not reported in any other organisms' RLRs yet. Another protein, EKC24604.1, is also interesting in its domain composition, which comprises of a CLECT instead of CTD in its C-terminal. We did not find any other protein with the similar domain architecture in the database of GenBank.

There is no basis to speculate that each of the putative oyster RLRs discussed here has a function in antiviral immunity at present, while the research on their function did emerge recently. The cDNA sequence of a RIG-I homolog in *C. gigas* (cgRIG-I) was cloned and studied [23]. The cgRIG-I sequence deduced from the cloned cDNA (KC702507) has a 100% identity with EKC31344.1 discussed in this study, it is apparent that cgRIG-I and the protein EKC31344.1 are actually derived from the same gene. Zhang's research showed that the mRNA level of oyster RIG-I in haemocytes is increased in response to poly (I: C) treatment. This result implied that the cgRLRs were involved in oyster's antiviral immunity, which is yet to be experimentally verified.

It is unusual that there are so many members of putative RLRs with a variety of domain architectures encoded in the genome of *C. gigas*. Comparatively, there is only one of RIG-I, MDA5 and LGP2 in the genome of human. We have found that in the genome of sea urchin, *Strongylocentrotus purpuratus*, a species of Echinodermata, the unusual high number of putative RLRs does also exist. We speculate that the unusual abundance of putative RLRs in certain organisms represents the state of halfway of the RLRs evolution or it is the result of the animals' adaptation to the specific environment.

Even though the genome sequence of *C. gigas* has not yet been assembled into individual chrosomes, it is can still be inferred that some of the putative RLRs genes are closed to each other in the genome according to their location at the sequence scaffold. It seems that some of the genes are tandem duplicated-at least partialy, and gives rise to the arrangement of the putative cgRLRs genes on scaffolds present in this study. In accord with this assumption, phylogenic analysis shows that those putative cgRLRs whose genes are vicinity in genome are closely related, such as EKC24603.1 and EKC24604.1, EKC38303.1 and EKC38304.1. The protein EKC34572.1 is not included in the phylogenetic analysis, while it is at the same scaffold and proximity to genes of EKC34571.1 and EKC34573.1, which are closed related in the phylogenic tree (figure 2). The phylogenic analysis of CARDs of the putative RLRs gave similar result (figure 3).

The number and location of introns in the putative cgRLRs genes are also investigated here, and in general, are not strictly conserved among the proteins studied. The varieties of cgRLRs are reflected not only by their different domain architecture, but also by the difference in structure of genes encoding them.

The different domains of RLRs play different function. RIG-I exists with a closed inactivation conformation in the absence of dsRNA binding [8]. The helicase and CTD domains recognize and bind viral RNA and release the CARDs, which then recruit and activate the signaling adaptor MAVS (IPS-1) [24]. MDA5 can cooperatively assemble into ATP-sensitive filaments on dsRNA and does not sequester its CARDs [9,10,25]. However, the CTD of MDA5 is required for cooperative filament assembly but not involved in RNA binding [10,26,27].The CARDs of MDA5 nucleate the assembly of MAVS into its active polymeric form, which can be promoted by K63-linked polyubiquitin chains [10,28,29]. LGP2 has similar helicase and CTD domains and recognize the dsRNA with the similar protein-RNA contacts as those of RIG-I and MDA5. While LGP2 lacks the tandem CARDs [30] and does not recruit MAVS or induce MAVS signalling [1]. Evaluating the domain architecture of all the putative cgRLRs, except those seemly are incomplete duplicates, the protein EKC38303.1 is unique that it contains only one N-terminal CARD. Whether or not the single CARD is sufficient to initiate downstream signaling is yet to be investigated further. Another protein, EKC24604.1, is unique in that it's a combination of a helicase domain and a CLECT (C-type lectin /C-type lectin-like) domain. Animal C-type lectins are involved in functions such as extracellular matrix organization, endocytosis, complement activation, pathogen recognition, and cell-cell interactions. It is possible that this kind of combination carry out the similar function as RLRs do in mammals.

In conclusion, it is surprising that so many putative cgRLRs with a variety of domain architectures encoded in the genome of *C. gigas*. The study of invertebrate immunity can give us some insight into the evolution of human innate immunity [31]. Elucidation of the putative function of RLRs in antivirus response of invertebrates may help to design novel antiviral strategies for virus control in Oyster which is an important economic species. It is valuable to research the functions of cgRLRs and their activities in oyster antiviral immunity.

References

[1] Reikine S, Nguyen J B and Modis Y 2014 Pattern recognition and signaling mechanisms of RIG-I and MDA5 *Front. Immunol.* **5** 342
[2] Gack M U 2014 Mechanisms of RIG-I-like receptor activation and manipulation by viral pathogens *J. Virol.* **88** 5213-6
[3] Rodriguez K R, Bruns A M and Horvath C M 2014 MDA5 and LGP2: Accomplices and antagonists of antiviral signal transduction *J. Virol.* **88** 8194-200
[4] Yoo J S, Kato H and Fujita T 2014 Sensing viral invasion by RIG-I like receptors *Curr. Opin. Microbiol.* **20** 131-8
[5] Fitzgerald M E, Rawling D C, Vela A and Pyle A M 2014 An evolving arsenal: Viral RNA detection by RIG-I-like receptors *Curr. Opin. Microbiol.* **20** 76-81
[6] Rawling D C and Pyle A M 2014 Parts, assembly and operation of the RIG-I family of motors *Curr. Opin. Struct Biol.* **25** 25-33

[7] Goubau D, *et al* 2014 Antiviral immunity via RIG-I-mediated recognition of RNA bearing 5'-diphosphates *Nature* **514** 372

[8] Kowalinski E, Lunardi T, McCarthy A A, Louber J, Brunel J, Grigorov B, Gerlier D and Cusack S 2011 Structural basis for the activation of innate immune pattern-recognition receptor RIG-I by viral RNA *Cell.* **147** 423-35

[9] Peisley A, Lin C, Wu B, Orme-Johnson M, Liu M, Walz T and Hur S 2011 Cooperative assembly and dynamic disassembly of MDA5 filaments for viral dsRNA recognition *Proc. Natl. Acad. Sci.USA.* **108** 21010-5

[10] Berke I C and Modis Y 2012 MDA5 cooperatively forms dimers and ATP-sensitive filaments upon binding double-stranded RNA *EMBO. J.* **31** 1714-26

[11] Talmage S C and Gobler C J 2010 Effects of past, present, and future ocean carbon dioxide concentrations on the growth and survival of larval shellfish *Proc. Natl. Acad. Sci. USA.* **107** 17246-51

[12] Walters E T and Moroz L L 2009 Molluscan memory of injury: evolutionary insights into chronic pain and neurological disorders *Brain. Behav. Evol.* **74** 206-18

[13] Zhang G, *et al* 2012 The oyster genome reveals stress adaptation and complexity of shell formation *Nature* **490** 49-54.

[14] Altschul S F, Gish W, Miller W, Myers E W and Lipman D J 1990 Basic local alignment search tool *J. Mol. Biol.* **215** 403-10

[15] Marchler-Bauer A, *et al* 2011 *CDD:* a Conserved Domain Database for the functional annotation of proteins. *Nucleic Acids Res.* **39** (Database issue) 225-9

[16] Edgar R C 2004 MUSCLE: A multiple sequence alignment method with reduced time and space complexity BMC. *Bioinformatics* **5** 113

[17] Tamura K, Stecher G, Peterson D, Filipski A and Kumar S 2013 MEGA6: Molecular evolutionary genetics analysis version 6.0 *Mol. Biol. Evol.* **30** 2725-9

[18] Langevin C, Aleksejeva E, Passoni G, Palha N, Levraud J P and Boudinot P 2013 The antiviral innate immune response in fish: evolution and conservation of the IFN system *J. Mol. Biol.* **425** 4904-20

[19] Pietretti D and Wiegertjes G F 2014 Ligand specificities of Toll-like receptors in fish: Indications from infection studies *Dev. Comp. Immunol.* **43** 205-22

[20] Wang P H, Weng S P and He J G 2014 Nucleic acid-induced antiviral immunity in invertebrates: An evolutionary perspective *Dev. Comp. Immunol.* **48** 291

[21] Green T J and Montagnani C 2013 Poly I: C induces a protective antiviral immune response in the Pacific oyster (Crassostrea gigas) against subsequent challenge with Ostreid herpesvirus (OsHV-1 muvar) *Fish. Shellfish. Immun.* **35** 382-8

[22] Canesi L, Betti M, Ciacci C, Citterio B, Pruzzo C and Gallo G 2003 Tyrosine kinase-mediated cell signalling in the activation of Mytilus hemocytes: Possible role of STAT-like proteins *Biol. Cell.* **95** 603-13

[23] Zhang Y, Yu F, Li J, Tong Y, Zhang Y and Yu Z 2014 The first invertebrate RIG-I-like receptor (RLR) homolog gene in the pacific oyster Crassostrea gigas *Fish. Shellfish. Immun.* **40** 466-71

[24] Zeng W, Sun L, Jiang X, Chen X, Hou F, Adhikari A, Xu M and Chen Z J 2010 Reconstitution of the RIG-I pathway reveals a signaling role of unanchored polyubiquitin chains in innate immunity *Cell.* **141** 315-30

[25] Berke I C, Yu X, Modis Y and Egelman E H 2012 MDA5 assembles into a polar helical filament on dsRNA *Proc. Natl. Acad. Sci. USA.* **109** 18437-41

[26] Li X, Lu C, Stewart M, Xu H, Strong R K, Igumenova T and Li P 2009 Structural basis of double-stranded RNA recognition by the RIG-I like receptor MDA5. *Arch. Biochem. Biophys.* **488** 23-33

[27] Takahasi K, *et al* 2009 Solution structures of cytosolic RNA sensor MDA5 and LGP2 C-terminal domains: identification of the RNA recognition loop in RIG-I-like receptors *J. Biol.*

Chem. **284** 17465-74

[28] Jiang X, Kinch L N, Brautigam C A, Chen X, Du F, Grishin N V and Chen Z J 2012 Ubiquitin-induced oligomerization of the RNA sensors RIG-I and MDA5 activates antiviral innate immune response *Immunity.* **36** 959-73

[29] Hou F, Sun L, Zheng H, Skaug B, Jiang Q X and Chen Z J 2011 MAVS forms functional prion-like aggregates to activate and propagate antiviral innate immune response *Cell.* **146** 448-61

[30] Yoneyama M, *et al* 2005 Shared and unique functions of the DExD/H-box helicases RIG-I, MDA5, and LGP2 in antiviral innate immunity *J. Immunol.* **175** 2851-8

[31] Ausubel F M 2005 Are innate immune signaling pathways in plants and animals conserved *Nat. Immunol.* **6** 973-9

Evaluation of the comprehensive potential for land reclamation in Huadian City

C Sun[1, 2, 3], Y Liu[1, 2, 3, 4], X M Tang[1, 2, 3] and Y M Ren[1, 2, 3]

[1] Beijing Research Center for Information Technology in Agriculture, Beijing 100097, China

[2] National Engineering Research Center for Information Technology in Agriculture, Beijing 100097, China

[3] Key Laboratory of Agri-informatics, Ministry of Agriculture, Beijing 100097, China

E-mail: Liuyu@nercita.org.cn

Abstract. In order to understand the potential for land reclamation in Huadian City, this research constructs an evaluation model for land reclamation potential based on the double restraints of ecological security and social economy to measure comprehensive potential for land reclamation that conforms to the level of regional development and ecologic protection objective. The research shows: (1) Under constraint conditions, the potential for land reclamation in the towns of Huadian City has been reduced in different degrees, with the change of potential in urban areas being much smaller than in rural area. Under the restraint of ecological protection, Jiapigou Town, Erdaodianzi Town, Hongshilazi Town and Changshan Town suffer the high restraint. In terms of restraint of social economy, Huashulinzi Town is subject to the high constraint. (2) During the 13th Five-Year Plan Period, through arable land reclamation, Huadian City will have 4,121.47 more hectares of arable land and save 2,410.82 hectares of construction land. The findings provide an approach to land reclamation potential evaluation and form a basis for the compiling of the plan for land reclamation during the 13th Five–Year Plan Period.

1. Introduction

As a basic task of land reclamation planning, potential evaluation is the basis of project site selection of land reclamation and scheduling, and constitutes the key technology to land reclamation planning [1]. In the context of regional overall development, land reclamation is more than reclamation of agricultural land, but the overall management of field, water, road, forest, and village [2]. In addition, ecological development requires that land reclamation should not only increase the amount of arable land and land capability, improve the function of the ecological environment, but also ensure urban and rural development as a whole and regional coordinated development [3]. The purpose of land reclamation potential evaluation is to identify the amount of the potential, level status, and the spatial distribution, so as to provide the basis for the scientific compiling of land renovation plan.

Currently, the addable effective cultivated land area after reclamations is used to represent the land reclamation potential. Zhang Z F *et al* [4, 5] explored the connotation and evaluation methods at home and abroad with regard to cultivated land reclamation potential and village residential land readjustment potential. Guo H Q *et al* [6] applied the multi-factor evaluation methods for cultivated land reclamation potential to a land reclamation project of Yanqing County. Yang W *et al* [7] used

GIS in their study of evaluation of the agricultural land reclamation potential in the Changshou District of Chongqing City. Wu L L *et al* [8] conducted a land reclamation potential evaluation (LRP) based on landscape principle. In short, GIS and other technologies are widely used in LRP evaluation, with the focus shifting from quantity potential to quality potential. Discussions on multi-factor comprehensive evaluation methods have begun, but most of the discussions are limited to cultivated land consolidation and rural residential land consolidation, among other individual potential evaluation, with a perfect LRP evaluation system yet to be established. LRP evaluation is the prerequisite for the scientific land reclamation planning and the study of LRP in the context of ecological safety and social economy development will help gain a deeper insight into the current development trend.

Huadian is a city with complex terrain, abounding in mountains and hills while lacking flat grounds. With the implementation of the national reforestation policy, and the increasing demand for construction land, the contradiction between the land supply and demand in the city is intensifying. It is therefore to perform land reclamation and tap into the land potential. In this study, the LRP of four types of land is explored, and a land reclamation potential evaluation model based on the double restraints of ecological safety and social economy is built to measure the overall LRP in Huadian City, to provide approach to the evaluation of LRP, and to provide reference for the land reclamation planning during the 13th Five-Year Period for the city.

2. Methodology

LRP refers to the potential under the context of certain natural conditions, regional social economic development level and ecological risk degrees, which requires modification of the traditional multi-factor evaluation model based on the restraint conditions.

$$F_i = \sum_{j=1}^{n} A_{ij} \times W_j \tag{1}$$

$$F_i = \frac{\sum_{j=1}^{n} A_{ij} \times W_j(A, T)}{T + 1} \tag{2}$$

$$T = \alpha T_e + \beta T_{s'} (\alpha + \beta = 1) \tag{3}$$

$$T_e = \sum_{i}^{m} W_E \times E_i \tag{4}$$

$$T_S = \sum_{i}^{l} W_s \times S_i \tag{5}$$

In the formula: F_i is the value of LRP of the ith evaluation unit; A_{ij} is the standardized value of the jth evaluation index of the ith evaluation unit; W_j is the weight of the jth evaluation index; W_j(A,T) means the effect weight on evaluation target of index j under constraint condition (T); T stands for constraint condition; T_e is the restrain index of ecological safety; T_s is the restrain index of social economy; α and β represent the effects of regional ecological safety and social economy on land reclamation respectively; E_i and S_i represent the constraint index factors of ecological safety and social economy; W_E and W_S stand for the index weight of the effects on ecological safety and social economy. In this study, $\alpha=\beta=0.5$. Based on the effects of index on LRP, leading indicators, positive constraint index and negative constraint index are defined. To offset the effect of measurement on data, extremum standardization is adopted for standardized processing of data, in addition to analytic hierarchy process for identifying the index weight. Drawing on the existing research results, a LRP evaluation index system based on ecological and social and economic constraints (see table 1).

Table 1. LRP evaluation index system and weight based on ecological and social and economic constraints.

Primary index and weight	Secondary index and its weights	Level 3 index and its weights	synthetic weight	orientation
Natural potential/0.7	Farmland reclamation potential/0.462	newly-increased farmland amount/0.166	0.116	leading
		newly-increased farmland coefficient/0.139	0.097	leading
		Capacity increased/0.139	0.096	leading
	Rural residential consolidation potential/0.220	Scale of rural residential consolidation/0.096	0.068	leading
		newly-increased farmland coefficient/0.083	0.061	leading
		Homestead idle rate/0.060	0.044	leading
	land reclamation potential/0.124	wasteland reclamation scale/0.057	0.030	leading
		newly-increased farmland coefficient/0.055	0.036	leading
	Unused agricultural land development potential/0.194	Unused agricultural land scale/0.123	0.090	leading
		newly-increased farmland coefficient/0.082	0.061	leading
Ecological security constraint/0.15	Water ecological security/0.397	Coverage of water source/0.397	0.060	positive constraint
	soil erosion risk/0.397	terrain slope/0.273	0.041	positive constraint
		Forest coverage/0.207	0.031	positive constraint
	Geological disaster risk/0.123	Geological disaster risk/0.123	0.018	positive constraint
Social economy constraint/0.15	Social development level/0.500	per-person cultivated land area/0.236	0.035	positive constraint
		Road network density/0.134	0.020	negative constraint
		Population density / 0.071	0.012	negative constraint
	Economic Feasibility/0.500	Per Capita GDP/0.250	0.037	negative constraint
		Rural per capita net income/0.169	0.027	negative constraint
		grain yield per hectare/0.140	0.020	negative constraint

- The evaluation index of natural potential consists of arable land reclamation potential, rural residential consolidation potential, mining wasteland reclamation potential and unused agricultural land development potential. (1) With the reduction of arable land, more importance is attached to the effect of land reclamation on the improvement of farmland quality, so the development potential for farmland quality should be included as a factor of natural potential. (2) Theoretical potential for rural residential consolidation is the difference between current rural residential area and that of the specified standard area, which can be represented by the scale of rural residential consolidation. Rural residential area can be turned into arable land after consolidation so as to relieve the land use conflict, so the newly-increased farmland coefficient should be regarded as an important indicator of potential, with the practical potential being represented by the idle rate of homestead. (3) Mining wasteland reclamation is mainly decided by the reclamation scale of wastelands which are left over by history or anticipated, with its newly-increased farmland coefficient deciding the contribution of reclamation to increased arable land. (4) The scale of unused agricultural land directly decides its development potential, with its development level subject to the regional ecological security conditions and natural landscape, and the potential value can be reflected by the newly-increased farmland coefficient.

- Ecological safety constraint index consists of water ecological security, soil erosion and geological disaster risk. (1) The land reclamation project which is too close to the water source may affect the regional ecological balance. So, coverage of water source is used to reflect the ecological security constraint factor of land reclamation, that is, the proportion of the area of land under evaluation within 500m of water source to the total area of land under evaluation. (2) Soil erosion, as one of the mountain ecological risk types, has a direct effect on the sustainability of land reclamation. The terrain slope and vegetation coverage are the important indicators of soil and water loss risk. (3) It is relatively hard to carry out land reclamation in areas where there is bigger risk of geological disaster. In this study, the area affected by geological disaster is used to reflect the regional geological disaster risk.
- Social economy constraint index includes social development level and economic growth. (1) Per-person cultivated land area decides the needs for land reclamation, and the bigger the per-person area, the lower the needs. Also, towns with developed economy and convenient transportation tend to attract population and expand itself, which indirectly increase the needs for land reclamation. So, per-person cultivated land area, road network density and population density are used to reflect the social constraint on the LRP. (2) GDP per capital, rural per capital net income, and grain yield per hectare are used to reflect the economic constraint on LRP.

3. Findings and analysis

Table 2 shows that LRP varies significantly among the towns of Huadian City, so does the LRP without constrains. Ecological constraint index indicates that Jiapigou Town is subject to the highest constraints, which may be partly due to its abounding in mountains and mines. By contrast, urban areas are exposed to fewer constraints. The towns subject to relatively or absolutely high socio-economic constraints account for 62.5% of total towns of the city. Huashulinzi Town is subject to the most constraints, with others exposed to more constraints. Without constraints, the LRP varies greatly among towns, with that of urban areas lower than that of towns. Jiapigou Town, Hongshilazi Town, Hengdaoheizi Town and Gongji Town enjoy a higher potential. With constraints, the changes in the potential in urban areas are less than that in towns.

Table 3 shows that a new round of land reclamation can add 4,121.47 hectares of arable land, of which 2,593.5 hectares are turned from the rearrangement of farmland, accounting for 62.92% of the total area of arable land. Rural construction land, after consolidation, offered 350.18 hectares arable land, accounting for 16.15%; reserved land suitable for cultivation contributed 512.27 hectares, accounting for 12.43%. The distribution of newly-increased farmland shows that farmland rearrangement, land reclamation and reserved land resources suitable for cultivation have made little contribution to the total area of newly-increased farmland. Jiapigou Town, however, relied mainly on land reclamation, Erdaodianzi Town on reserved land suitable for cultivation, and other towns on the rearrangement of farmland in their newly increasing of arable land.

Table 3 shows that Huadian City can save 2,410.82 hectares of construction land, with 387.12 hectares coming from the rearrangement of rural construction land, accounting for 16.06%, and 2,023.70 hectares from the rearrangement of industrial and mining land for construction purpose, making up 83.94%. The scale of rearranged land for rural construction differs greatly from that for industrial and mining construction in the towns. The land rearranged for rural construction is mainly located in Yongji Sub-District, Jiapigou Town, Hongshilazi Town and Gongji Town, occupying 59.65% of the total rearranged land; the land rearranged for industrial and mining construction is mainly located in Yongji Sub-District, Jiapigou Town, Hongshilazi Town and Huajiao Town, accounting for 76.30% of the total rearranged land.

Table 2. Results of land potential evaluation in Huadian.

Township name	Ecological constraint index		Socioeconomic constraint index		Unrestricted potential value		Potential value under constraints	
	sort	Constraint level	sort	Constraint level	sort	Constraint level	sort	Constraint level
Minghua	14	Low	16	Low	13	Low	15	Low
Yongji	12	Low	11	General	12	General	12	General
Shengli	13	Low	12	General	16	Low	16	Low
Xinhua	15	Low	13	General	15	Low	14	Low
Qixin	16	Low	14	General	14	Low	13	Low
Jiapigou town	1	High	9	Higher	3	High	3	General
Erdaodianzi	3	High	8	Higher	5	General	10	General
Hongshilazi	2	High	15	General	1	High	2	General
Badaohezi	6	Higher	10	Higher	9	General	9	General
Changshan	4	High	4	Higher	11	General	11	General
Huajiao	11	General	7	Higher	7	General	6	General
Hengdaohezi	8	Higher	3	Higher	4	High	4	General
Jinsha town	10	General	2	Higher	10	General	8	General
Huashulinzi	9	Higher	1	High	6	General	5	General
Gongji town	7	Higher	6	Higher	2	High	1	General
Sumigou	5	Higher	5	Higher	8	General	7	General

Table 3. Supplementary cultivated land potential form land consolidation and intensive construction land saving potential in Hudian City.

Township name	Supplementary cultivated land potential /hectare					Intensive construction land saving potential/hectare		
	Agricultural land consolidation	Rural construction land consolidation	Land reclamation	Reserve land resources development	Total	Rural construction land consolidation	Urban and rural construction land consolidation	Total
Minghua	0.57	0	0.63	0.00	1.2	1.29	86.08	87.37
Yongji	29.54	40.86	12.44	34.68	117.52	45.41	270.18	315.59
Shengli	0	0	0.00	0.00	0	0	31.8	31.8
Xinhua	0.04	0	0.00	0.00	0.04	0	0	0
Qixin	0.2	0	0.00	0.00	0.2	0	0	0
Jiapigou	98.45	47.67	342.02	87.65	575.79	51.81	568.1	619.91
Erdaodianzi	62.41	11.30	23.60	173.18	270.49	11.77	80.3	92.07
Hongshilazi	456.23	52.04	67.09	53.24	628.6	56.57	488.84	545.41

Badaohezi	218.14	12.00	52.74	12.28	295.16	12.76	181.78	194.54
Changshan	101.31	15.17	23.65	41.75	181.88	16.49	58.18	74.67
Huajiao	260.88	19.76	75.44	22.12	378.2	21.71	216.91	238.62
Hengdaohezi	267.86	3.30	6.29	27.62	305.07	3.67	0	3.67
Jinsha town	245.87	5.96	16.35	21.59	289.77	6.69	8.12	14.81
Huashulinzi	164	50.76	9.12	14.01	237.89	54	4.97	58.97
Gongji town	521.18	66.33	21.49	11.60	620.6	77.14	10.42	87.56
Sumigou	166.82	25.03	14.66	12.54	219.05	27.81	18	45.81
Total	2593.5	350.18	665.52	512.27	4121.47	387.12	2023.7	2410.82

4. Discussion and conclusion

Land reclamation is a necessary step towards the increase of utilization rate of arable land. In this study, the dual constraints of social economy and ecological security are introduced for a quantitative evaluation of the overall LRP in Huadian under the constraints of ecological security and social economy, based on an analytic hierarchy process and evaluation model with constraints. Jiapigou Town, Erdaodianzi Town, Hongshilazi Town, and Changshan Town are exposed to the high constraints of ecological security. Huashulinzi Town is subject to the high constraints of social economy. The findings show that social economy and ecological security have an important effect on LRP, the advanced the economy and the safer the ecological environment, the higher the potential, and vice versa. During the 13th Five-Year Plan Period, land reclamation can bring about newly-increased farmland of 4,121.47 hectares, and save construction land of 2,410.82 hectares. At last, the model under constraints, though reflecting to some extent the LRP in areas in question and fitting the local conditions, calls for further research to see whether it can apply to other areas. In addition, the evaluation indexes of the model of constraints established in this study are limited in number and need to be further improved.

Acknowledgments
This research is supported by the National Natural Science Foundation of China (41401193).

References

[1] Liu X L and Zhang W 2014 Study on the potential measurement for land renovation plan at the county level-A case of Shenmucounty *J. Arid Land Resour. Environ.* **28** 33-8
[2] Yan J M, Xia F Z and Ma M 2016 Strategy orientations of transformation development: Land consolidation in the new period of China *Chin. Land Sci.* **30** 3-10
[3] Li Q, Dai L and Zhu Q 2014 Ecological connectivity changes and its pattern optimization during landconsolidation based on minimal accumulative resistance model *Scientia Geogr. Sin.* **34** 733-9
[4] Zhang Z F, Wang X L and Guo B Y 2015 Effects on rural residential land consolidation in Metropolitan Suburb *China Land Sci.* **29** 18-24
[5] Zhang Z F 2012 Criteria and assessment for sustainability of land consolidation *Transac. Chin. Soc. Agri. Eng.* **28** 1-7
[6] Guo H Q, Wang L, Fan J M, Zhang Y and Yang H 2006 Multi-factor comprehensive evaluation model for the potential of arable land consolidation for Yanqing District in Beijing *Transac. Chin. Soc. Agri. Eng.* **22** 83-86
[7] Yang W, Xie D T, Liao H P, Pan Z and Zhu L 2013 Analysis of consolidation potential of agricultural land based onconstruction mode of high-standard basic farmland *Transac. Chin.*

Soc. Agri. Eng. **29** 219-29
[8] Wu L L, Luo J P and Li M 2010 Evaluation method of land-scaled consolidation potential based on landscape pattern principle *Transac. Chin. Soc. Agri. Eng.* **26** 300-6

Characterization of pre-curing behavior of urea-formaldehyde resin affected by different temperatures

Y Z Chen[1,2] and H Xiao[1,2,3]

[1] College of Forestry, Sichuan Agricultural University, Chengdu, 611130, Sichuan, China
[2] Key laboratory of wood industry and furniture engineering, Sichuan Agricultural University, Chengdu, 611130, Sichuan, China

E-mail: xh_b70405@126.com

Abstract. Differential scanning calorimetry (DSC) was applied to investigate the thermal behavior of pre-curing urea-formaldehyde (UF) resin as affected by different temperatures; the activation energies (Ea) in different stages during the pre-curing course were analyzed by Kissinger method. The results indicated that the curing time was decreased with temperature increasing. Under isothermal temperature, the DSC curves of pre-curing UF resin shifted to shorter time, the conversion increased with increasing temperature. In dynamic scanning, with the pre-curing degree increasing, the DSC curves shifted to lower temperature, while both the onset and peak temperature decreased due to the cross-linking degree increased. The Ea and Z value decreased obviously firstly due to the concentration and reactivity of the reactants increased with water evaporation in the first stage, and then the increasing cross-linking degree and the formation of the network structure lead to both the Ea and Z value increasing in the second stage, indicating that the pre-curing behavior appeared.

1. Introduction

Urea-formaldehyde (UF) resin is a dominant thermosetting wood adhesive for the production of wood composite boards, such as medium density fiberboard (MDF), particleboard (PB) and hardwood plywood for its good performance.

Initially, mechanical blending was the main method to apply UF resin to dried fibers in MDF manufacture, but it was soon replaced by blowline blending method due to the resin distribution was much more uniformly and resin spots would not remained on the board surface. In the drying blow-line, the environment of curing agents, high temperature, and humidity would induce the occurrence of cross-linking reactions and pre-curing behavior, resulting in low efficiency of the resin and high cost.

Thermosetting resins changed from liquid state to viscous stage then gelation and vitrification till cured for the tight network formation [1]. Agglomeration of resin appeared to occur during heating treatments and storing course [2], the formation of methylenes and dimethylene ethers were the main reaction, the active functional groups and molecular mobility of aged resin were decreased resulted in lowing of cure rate [3]. Siimer *et al* [4] evaluated the cure process of UF resins and the thermal stability of cured resin with TG-DTA method; it was found that the exothermic curing temperature shifted to higher temperature with decreasing of the content of hydroxymethyl groups during storage. With Dynamic Mechanical Analysis (DMA), Young No *et al* [5] found that the gel times shortened and cure slope of the vitrification stage decreased with elapsed time increasing, due to the deleterious effects of polycondensation.

However, few researches have focused on the pre-curing behavior of the UF resin, and the correlativity between the temperature, elapsed time and the pre-curing behavior was unknown. In this study, the thermal behavior of the pre-curing UF resin affected by different temperature was discussed, and the pre-curing kinetics during the whole pre-curing course were also analyzed.

2. Materials and methods

2.1. Materials
The commercial Urea-formaldehyde resin powder adhesive (PrefereTM 4405) was purchased from DYNEA ASA Chemical Co. Ltd (Shanghai, China). Based on the total resin weight, a 1% ammonium chloride (NH_4Cl) (25 % aqueous solution) was added as a curing agent. The UF resin with the solid content of 40 %, 50 %, 60% was prepared by dissolving the powder with the distilled water.

2.2. Samples Preparation
About 1g of UF and ammonium chloride mixture was placed in aluminum tube with the dimension of 96.6 mm × 160 mm×0.7 mm (length ×diameter× wall thickness). When the temperature of the glycerin reached 80, 100, 120 and 150°C, the tubes were put into oil bath with 2 cm below the oil level. Once reaching the setting time, the tube was taken out and then immersed into the liquid nitrogen.

2.3. Curing time measurement
According to China Industry Standard (GB/T 14074-2006), about 10g UF resin by adding 1 wt% NH_4Cl (25 wt% solution) placed into a test tube with the dimension of 150 mm × 25 mm (length×diameter), and then the tubes were put into the oil bath with stirring manually with three replication for each sample. Once the gelation was occurred, the measurements were finished.

2.4. Thermal behavior measurement

2.4.1. Isothermal DSC analysis
About 5 mg of samples were placed between an aluminum cup and lid, then sealed tightly. The pans were placed into sealed cell of DSC (PE diamond) instrument and pressurized to 2.0 MPa with nitrogen. The setting isothermal temperature was 80°C, 100°C, 120°C and 150°C, the heating rate was 200°C/min.

2.4.2. Dynamic DSC analysis
A dynamic DSC scanning (TA Q2000) equipped with TA universal 2000 analysis software was used to evaluate the non-isothermal behavior. About 2 mg of the UF resins was weighted in high-pressure crucible prior to testing with the testing temperature range from 40 to 150°C and the heating rate of 10, 15, 20 and 25°C/min. The non-isothermal testing temperature was 80°C, 100°C, 120°C and 150°C.

2.5. Activation energy analysis
A single dynamic scan method was used to calculate the activation energy (E_a) based on the *n*th-order kinetics expressed by the following Kissinger equation (as shown in equation (1)):

$$-\ln\left(\frac{\beta}{T_p^{\ 2}}\right) = \frac{E_a}{RT_p} - \ln\left(\frac{ZR}{E}\right) \tag{1}$$

On the basis of equation (1), the E_a (activation energy, kJ/mol) can be obtained from the the slope of the straight line between $-\ln(\beta/T_p^2)$ and $1/T_p$, while the Z (pre-exponential factor, 1/s) can be calculated from the y-intercept refers to $-\ln(Z\times R/E_a)$.

3. Results and discussion

3.1. Effect of the temperature on the curing time

As can be seen in figure 1, curing temperature showed significant effect on the curing time. The curing time of resin with the solid content of 40%, 50% and 60% decreased with increasing the temperature. When the solid content was 50%, the curing time was 193 s for 80°C, 96 s for 120°C and 47 s for 180°C. It is generally believed that under the effect of temperature in drying pipe, the solid content of resin was increased due to the evaporation of free water. On the basis of hydrogen bonds, the changing amount and compatibility of the water, which was regarded as active component to form physical network, would lead to different sedimentation conditions during formation of network of different densities [6].

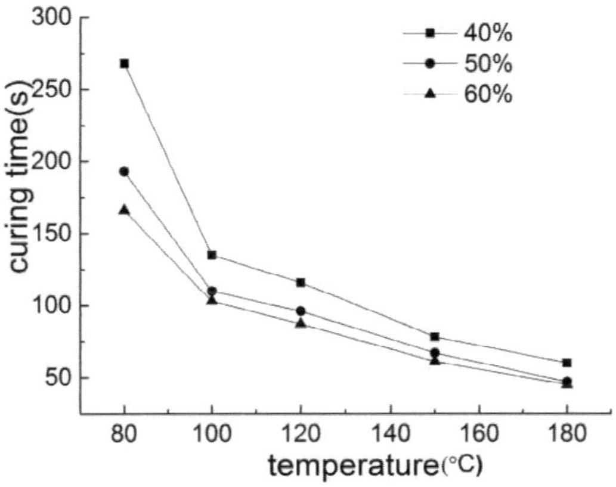

Figure 1. The curing time of UF resins with different temperature.

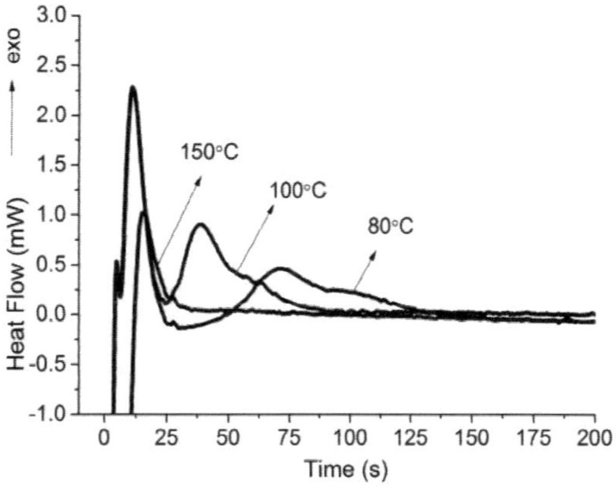

Figure 2. Isothermal DSC curves of UF resin under different temperature.

3.2. Thermal behavior of the pre-curing UF resin

3.2.1. Thermal behavior with isothermal condition

The results of thermal behavior of UF resins under the isothermal condition at different temperature were shown in figure 2 and table 1. As shown, the whole curing curves were shifted to shorter time with increasing temperature, and both the values of heat and areas of curing peak were similar between 80°C and 100°C. When temperature increased to 150°C, the values of heat and peak area were lower than that of 80°C and 100°C, this deviation was mainly ascribed to fast curing under 150°C, the section of initial curing curve was overlapped with the part of adjustment of instrument (the adjustment time of isothermal DSC was about 20 s), and the true onset and peak time were hided. As shown in figure 3, at the same time, the conversion values of UF resin were higher with increasing temperature.

Table 1. The isothermal DSC parameters of UF resin.

Temperature (°C)	Onset time (s)	Peak time (s)	End time (s)	Heat (J/g)	Peak area (mJ)
80	51.12	70.02	122.22	20.027	29.039
100	28.26	39.18	54.72	17.832	28.695
150	12.42	15.6	22.98	4.098	7.376

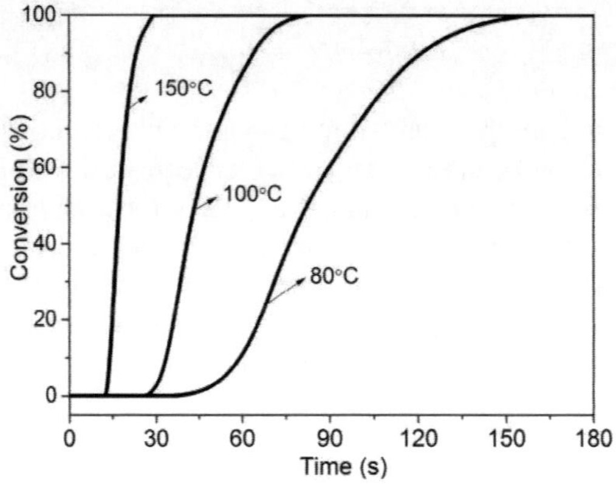

Figure 3. The conversion of UF resin under different temperature.

3.2.2. Thermal behavior with non-isothermal condition

Figure 4 shows typical DSC curves of UF resins with different pre-curing time under the temperature of 100°C and 150°C. It can be seen that most curves showed more than one exothermal peak (i.e., the control and 10 s in the figure 4(a)), this phenomenon could be attributed to the specific conditions of synthesis formulas of the high formaldehyde [7] content, the different NaOH content [8] and urea component [9,10]. It can be seen that the exothermic peaks shifted to lower temperature and became smaller with the pre-curing time prolonging.

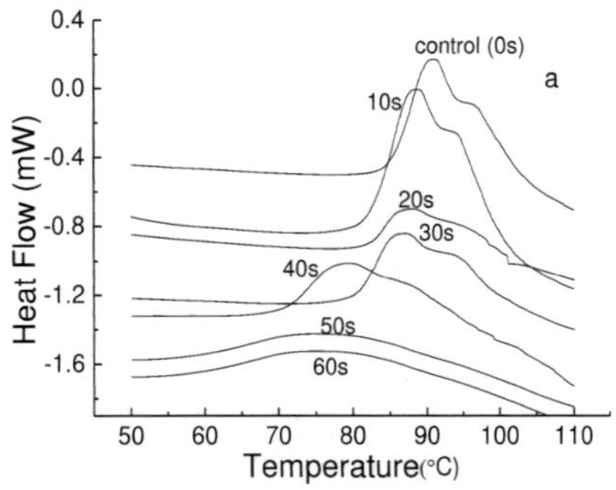

(a)

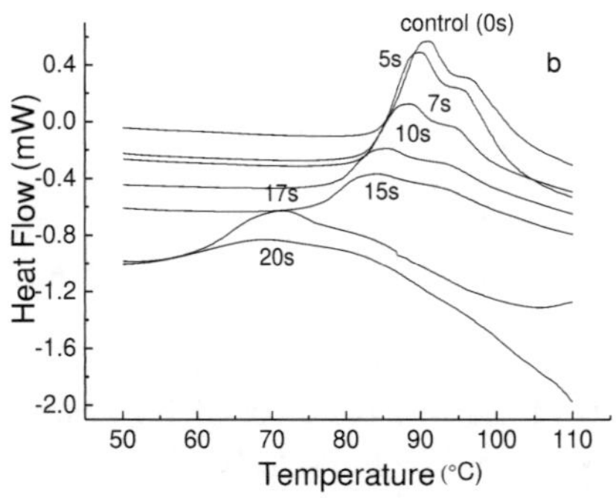

(b)

Figure 4. The DSC curves of UF resins with different pre-curing time ((a) 100°C, (b) 150°C).

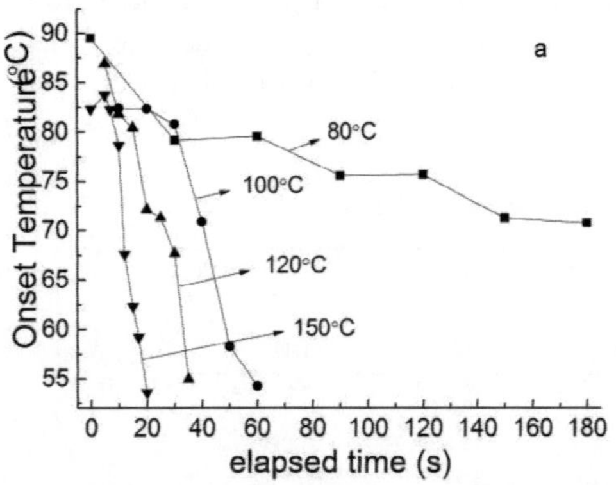

(a)

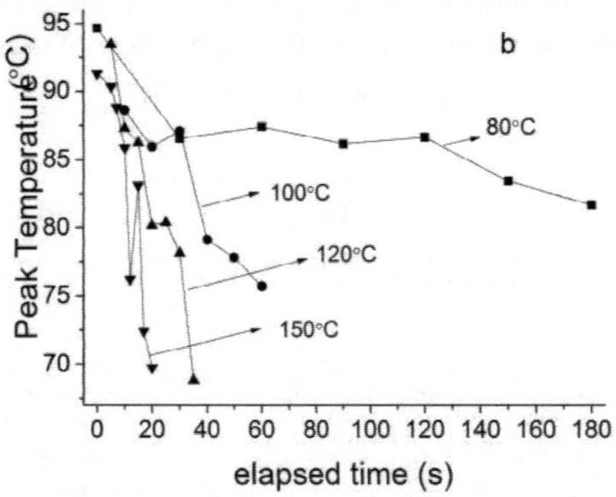

(b)

Figure 5. The onset and peak temperature of DSC with different temperature and elapsed time ((a) onset temperature, (b) peak temperature).

According to DSC curves, the onset and peak temperature was the time point when curing reaction began and the conversion rate reached a maximum during a dynamic scan, respectively [11]. As shown in figure 5, the temperatures of the onset and peak decreased remarkably with time elapsing. It can be concluded that the curing temperature were higher, the polycondensation rate of the active functional groups became faster leading to the cross-linking degree of the pre-curing UF increased.

3.3. Curing Kinetic of pre-curing UF resin under different temperature

The results of the E_a, Z and corresponding regression coefficient (R_a^2) calculated by the Kissinger method were shown in table 2 and figure 6.

Table 2. The activation energy of pre-curing UF resins.

	Pre-curing time (s)	heating rate (°C /min) β	Peak Temperature (°C)	LN (β/T_P^2)	$1/T_P$	E_a (kJ/mol)	Z-pre exponential	R_a^2
control	0	10	94.62	-9.512	0.0027	100.24	1.61×10^{11}	0.946
		15	98.72	-9.129	0.0027			
		20	100.21	-8.849	0.0027			
		25	104.77	-8.650	0.0026			
80°C	30	10	86.55	-9.468	0.0028	53.97	3.31×10^{4}	0.917
		15	97.45	-9.122	0.0027			
		20	100.37	-8.850	0.0027			
		25	102.63	-8.639	0.0027			
	90	10	86.17	-9.466	0.0028	49.89	9.67×10^{5}	0.840
		15	89.53	-9.079	0.0028			
		20	93.72	-8.814	0.0027			
		25	103.30	-8.643	0.0027			
	150	5	76.46	-10.104	0.0029	84.98	2.27×10^{9}	0.930
		10	81.70	-9.441	0.0028			
		15	90.14	-9.082	0.0028			
		20	90.31	-8.796	0.0028			
100°C	10	10	89.21	-9.483	0.0028	96.13	6.53×10^{10}	0.967
		15	92.66	-9.096	0.0027			
		20	97.40	-8.834	0.0027			
		25	98.36	-8.616	0.0027			
	30	10	83.79	-9.453	0.0028	65.17	2.05×10^{6}	0.977
		15	91.63	-9.091	0.0027			
		20	95.01	-8.821	0.0027			
		25	97.74	-8.613	0.0027			
	50	10	77.81	-9.419	0.0028	167.65	1.75×10^{22}	0.855
		15	78.14	-9.015	0.0028			
		20	80.63	-8.742	0.0028			
		25	82.46	-8.529	0.0028			
120°C	10	10	87.30	-9.472	0.0028	65.64	1.97×10^{6}	0.947

		15	93.73	-9.102	0.0027			
		20	99.77	-8.847	0.0027			
		25	100.08	-8.626	0.0027			
	20	10	80.15	-9.432	0.0028	65.47	3.41×10^{6}	0.947
		15	84.34	-9.050	0.0028			
		20	85.47	-8.769	0.0028			
		25	93.40	-8.589	0.0027			
	30	10	76.67	-9.412	0.0029	67.37	7.30×10^{6}	0.900
		15	85.68	-9.058	0.0028			
		20	86.87	-8.777	0.0028			
		25	88.68	-8.563	0.0028			
150°C	5	10	90.36	-9.489	0.0028	96.04	5.34×10^{10}	0.884
		15	96.01	-9.114	0.0027			
		20	98.80	-8.842	0.0027			
		25	100.00	-8.625	0.0027			
	10	10	85.36	-9.461	0.0028	66.45	3.12×10^{6}	0.944
		15	89.63	-9.080	0.0028			
		20	96.78	-8.831	0.0027			
		25	97.67	-8.613	0.0027			
	15	10	83.38	-9.450	0.0028	120.02	4.26×10^{14}	0.990
		15	87.44	-9.067	0.0028			
		20	89.62	-8.792	0.0028			
		25	91.14	-8.577	0.0027			

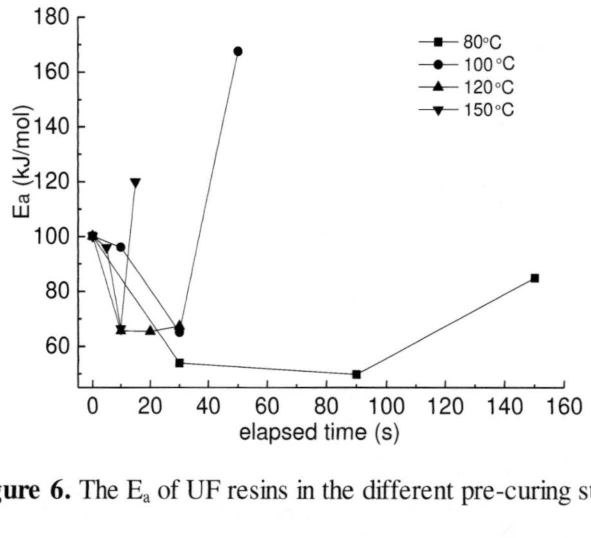

Figure 6. The E_a of UF resins in the different pre-curing stages.

As shown in figure 6, the changes in activate energy (E_a) were decreased initially, then ascended dramatically. On the basis of collision theory, the Z value is on behalf of the number of active sites and collision possibilities [12], the variation tendency of Z values was in accordance with that of the values of E_a.

The variations of the E_a and Z values were obviously decreased firstly and then increased, however, the time nodes of each stage were unsynchronized with different temperature, in other words, the higher of the curing temperature, the shorter of the time course in which both the values of E_a and Z begin to decrease and then increase.

This result was quite reasonable, (1) the collision chances of molecules were reduced in the early stage of the curing process due to the abundant water in the fresh resin. When resin was heated under isothermal conditions, the constant effect of heat promoted evaporation of water, and the concentration of the reactants increased resulting in the decreasing of active sites and increasing of cross-linking degree. This was the cause of the decreasing of E_a and Z in the first stage. (2) In the second stage, the water was locked in the macro-molecular structure due to the formation of the super-cluster and the tight network [3,13], and then more energy was required to promote the water evaporation, which lead to the E_a and Z value increased and the reaction rate decreased along with pre-curing time increasing. And the increase of E_a and Z value make it possible to judge the occurrence and the degree of the pre-curing behavior. These results were in a good agreement of the conclusion in our previous researches [14].

In general, activation energy has a positive relationship with the pre-curing behavior of the UF resin; these results indicated that pre-curing behavior of the UF resin could be predicted by evaluating of the variation of the activation energy by DSC.

4. Conclusions

- The temperature had strong effect on the curing time. The resin curing action was subjected to double effects of drying temperature and solid content due to the evaporation of free water in the drying process.
- Under the isothermal temperature, with temperature increasing, the whole curing curves were shifted to shorter time, both heat and peak areas were similar at 80°C and 100°C, while these two values were higher than that of 150°C due to overlapped of curing curves.
- With the dynamic scanning, the whole DSC curves of UF resin shifted to lower temperature and the onset temperature and peak temperature decreased remarkable with increasing of the pre-curing degree. Meanwhile, the value of Ea and Z decreased initially and then increased due to the evaporation of water and the formation of the in tight network of UF resin. These results indicated that pre-curing behavior of the UF resin could be predicted by evaluating of the variation of the activation energy by DSC. In short, lowing the drying temperature is a proper method to slow down the pre-curing reaction of UF resin.

Acknowledgments

This work was supported by Undergraduate Innovative Training Program of Sichuan Agricultural University (Project No. 04054631) and Research Interest in Training Program of Sichuan Agricultural University (Project No. 04054503).

References

[1] Babayevsky P G and Gillham J K 1973 Epoxy thermosetting systems: Dynamic mechanical analysis of the reactions of aromatic diamines with the diglycidyl ether of bisphenol *A J. Appl. Polym. Sci.* **17(7)** 2067-88

[2] Kim M G, No B Y, Lee S M and Nieh W L 2003 Examination of selected synthesis and room-temperature storage parameters for wood adhesive-type urea–formaldehyde resins by 13C-NMR spectroscopy. V *J. Appl. Polym. Sci.* **89(7)** 1896-917

[3] Siimer K, Kaljuvee T, Christjanson P and Pehk T 2005 Changes in curing behaviour of aminoresins during storage *J. Therm. Anal. Calorim.* **80(1)** 123-30

[4] Siimer K, Kaljuvee T and Christjanson P 2003 Thermal behaviour of urea-formaldehyde resins during curing *J Therm Anal Calorim.* **72(2)** 607-17

[5] Young No B and Kim M G 2005 Curing of low level melamine-modified urea-formaldehyde particleboard binder resins studied with dynamic mechanical analysis (DMA) *J. Appl. Polym.*

Sci. **97(1)** 377-89
[6] Park B D, Riedl B, Hsu E W and Shields J 1999 Differential scanning calorimetry of phenol–formaldehyde resins cure-accelerated by carbonates *Polymer.* **40(7)** 1689-99
[7] Christiansen A W and Gollob L 1985 Differential scanning calorimetry of phenol–formaldehyde resols *J. Appl. Polym. Sci.* **30(6)** 2279-89
[8] Tonge L Y, Hodgkin J, Blicblau A S and Collins P J 2001 Effects of initial phenol-formaldehyde (PF) reaction products on the curing properties of PF resin *J. Therm. Anal. Calorim.* **64(2)** 721-30
[9] Zhao C, Pizzi A, and Garnier S 1999 Fast advancement and hardening acceleration of low - condensation alkaline PF resins by esters and copolymerized urea *J. Appl. Polym. Sci.* **74(2)** 359-78
[10] He G and Riedl B 2003 Phenol-urea-formaldehyde cocondensed resol resins: Their synthesis, curing kinetics, and network properties *J. Polym. Sci. Part B: Polym. Phy.* **41(16)** 1929-38
[11] Xing C, Deng J, Zhang S Y, Riedl B and Cloutier A 2005 Differential scanning calorimetry characterization of urea–formaldehyde resin curing behavior as affected by less desirable wood material and catalyst content *J. Appl. Polym. Sci.* **98(5)** 2027-32
[12] Jiang J X, Yang Y L, Li C and Li J Z 2011 Effect of three boron flame retardants on thermal curing behavior of urea formaldehyde resin *J. Therm. Anal. Calorim.* **105(1)** 223-8
[13] Pizzi A, George B, Zanetti M and Meausoone P J 2005 Rheometry of aging of colloidal melamine–urea–formaldehyde polycondensates *J. Appl. Polym. Sci.* **96(3)** 655-9
[14] Chen Y Z, Fan D B, Qin T F and Chu F X 2015 Assessment of thermal behavior of pre-curing UF resin with different heat treatment *J. Adhes. Sci. Technol.* **29(18)** 1902-10

IOP Conf. Series: Earth and Environmental Science **77** (2017) 012008 doi:10.1088/1755-1315/77/1/012008

Spatiotemporal dynamics of climate and agricultural landscape patterns in Xinjiang, China (2000-2015)

T Huang[1], Q Y Du[1] and J Q Zou[2, 3]

[1] School of Resources and Environmental Science, Wuhan University, Wuhan 430079, China

[2] Institute of Agricultural Resources and Regional Planning, Chinese Academy of Agricultural Sciences, Beijing 100081, China

E-mail: zoujinqiu@caas.cn/40271472@qq.com

Abstract. This study integrated remote sensing, geographic information system, landscape ecology, and spatial analysis, and is based on meteorological data and four remote sensing images (2000-2015). The aim of this study was to analyse climate trends and spatiotemporal changes in agricultural landscape patterns, and calculate agricultural ecosystem service value (AESV) in Xinjiang, China. In addition, correlation analysis further enabled the quantification of agricultural landscape pattern changes in response to climate change. Results showed that temperature and precipitation registered a fluctuating ascending trend in most areas of Xinjiang. Landscape diversity and fragmentation increased during the same period. AESV showed a trend of gradual increase, the waste disposal and the water retention are the most important ecological functions. Besides, agricultural landscape pattern significantly correlated to climate and the effect of precipitation on agricultural landscape patterns has been greater than the effect of temperature during 2000-2015.

1. Introduction

Sustainable development is a key factor that characterizes modern agricultural practices [1, 2], and ecological agriculture is the foundation for the sustainable development of agriculture. Thus, ecological agriculture has gradually become a relevant research topic [3]. An agricultural landscape is formed by the interaction of a patch, which is one of the most widely distributed landscape types in the world. Indeed, because of the effects of the natural environment and human activity, agricultural landscape structures are constantly changing, and tremendous efforts have been made to understand the driving forces [4] underlying agricultural landscape patterns, and their ecological environmental effects [5] and so on. Recently, climate change characterized by global warming has become one of the most important environmental problems in the world [6]. Climate change has not only a significant impact on agricultural production [7, 8], but also leads to alterations in the composition, structure, and function of agricultural ecosystems, as well as changes in biodiversity [9]. Studies on the dynamic changes of agricultural landscapes as a result of the background effects of climate warming will make great contribution to farmers and science community [6]. Agricultural ecosystems play important roles in water conservation, gas regulation as well as in many other dynamics [10], and agricultural ecosystem service value (AESV) is one embodiment of agricultural ecological effectiveness. In previous work, Costanzo [11] has had the most far-reaching impact on the quantification of values for global ecosystem services.

IOP Conf. Series: Earth and Environmental Science **77** (2017) 012008 doi:10.1088/1755-1315/77/1/012008

The goals of this study are to: analyze climate trends and spatiotemporal changes in agricultural landscape patterns during 2000-20215; calculate AESV of Xinjiang; and further quantify agricultural landscape pattern changes in response to climate change.

2. Materials and methods

2.1. Study area
Xinjiang is located in the northwest of China (73°32'–96°21'E, 34°22'–49°33'N), and encompasses about one-sixth of total Chinese land area. Indeed, large variations in temperature, reduced precipitation, and a dry climate with the annual precipitation of only 167.1 mm and the annual average temperature of 10°C are observed [12]. In Xinjiang, agricultural production predominates, especially cotton production, and the area of cultivated land is the largest in China. As the core area of the Silk Road Economic Belt, the agricultural development of this region is hugely important to economy [13], but because this region comprises arid and semi-arid inland environments, conditions for the development of ecological agriculture are relatively poor [14].

2.2. Data sources and analysis
Four periods of Chinese remote sensing land use data for 2000, 2005, 2010, and 2015 were collected from the Data Center for Resources and Environmental Sciences, Chinese Academy of Sciences (http://www.resdc.cn). These data were visually interpreted from Landsat TM/ETM remote sensing images and comprise six types. In this study, agricultural landscape is defined as including farmland, forest, grassland, and water.

Meteorological data (2000-2015) were retrieved from National Meteorological Administration, including meteorological stations (54), which have data of daily precipitation, and daily average temperature. The Kriging interpolation, which takes sufficiently the size, shape and spatial position of the certain points into account, especially the correlation of spatial position between the unknown points [15], was calculated to obtain a spatial distribution map of extreme highest temperature (EHT), extreme lowest temperature (ELT), annual average temperature (AAT) and annual precipitation (AP) via ArcGIS.

Grain crop data were extracted from the 'Xinjiang Statistical Yearbook' and 'National Agricultural Product Cost and Income Data Compilation'. The major crops in Xinjiang include rice (6%), wheat (50%), corn (40%), and barley (2%).

2.3. Landscape metric analysis
Landscape metrics can be used to effectively reflect information on landscape patterns from structural composition and spatial distribution characteristics. In order to reflect properties of a whole landscape, highlight important features and avoid redundant information, metrics including total landscape area (TA), patch density (PD), Shannon's diversity index (SHDI), contagion index (CONTAG) and aggregation index (AI) for this study were selected, Landscape metrics were calculated using Frag stats 4.2 software.

The equation for the calculation of change in agricultural landscape pattern was as follows:

$$R = \frac{V_{t+n} - V_t}{V_t} \times 100\% \tag{1}$$

Where R is the change rate of agricultural landscape metrics between the year $t+n$ and year t, V_{t+n} is the landscape pattern metric for the year $t+n$, V_t is the landscape pattern metric for the year t, n is the time interval.

2.4. Assessment of AESV
According to Xie *et al* [16], the equation used for assessing ecosystem service values is as follows:

$$E_a = \frac{1}{7}\sum_{i=1}^{n}\frac{m_i p_i q_i}{M} \quad (2)$$

Where E_a is the value of the unit equivalent factor, p_i is the national average price for crop i, q_i is the unit yield for crop i, m_i is the growing area for crop i, and M is the total planting area for all crops. The ecosystem service values for different land use types were calculated using the following equation:

$$AESV = \sum(A_k \times VC_k) \quad (3)$$

Where $AESV$ denotes the ecological service value, A_k represents the area of land use of k type, while VC_k is the ecological service coefficient of this land use type.

3. Results and discussion

3.1. Climate change

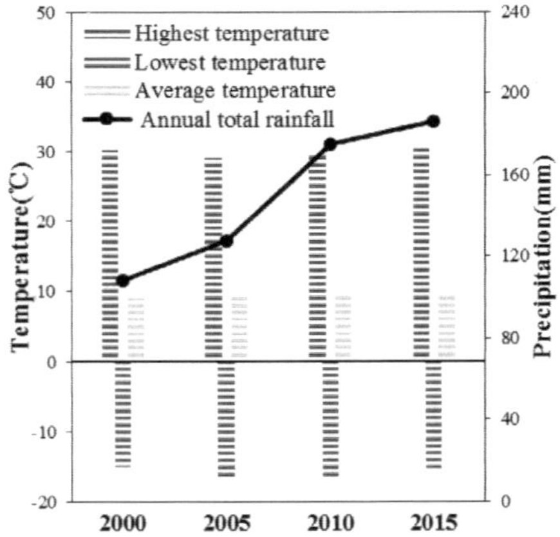

Figure 1. Climate of Xinjiang (2000-2015).

Table 1. Changes of EHT, ELT, AAT and AP in in each region of Xinjiang (units are °C and mm).

	EHT			ELT			AAT			AP		
	T1[a]	T2	T3	T1	T2	T3	T1	T2	T3	T1	T2	T3
Aksu	-1.8	0.8	1.4	-2.8	0.4	1.8	-0.5	0.8	-0.5	46	69.2	-16.4
Altay	-0.9	0.5	0	-2.1	0.2	1.5	0.3	-0.9	1.2	-4.1	75.4	-56.8
Korla	-1.3	0.9	1.6	-1.3	-0.6	2	0.3	0.3	0.3	5.9	11.7	32.7
Bole	-0.1	-0.8	2.7	-4	1.4	1.9	-0.6	0.2	0.1	46.3	109	-53.4
Changji	-1.1	0.1	1.1	-2.3	-2.3	4	0.6	-0.5	0.8	-14.4	75.7	19.9
Hami	0.3	1.0	-1.3	-1.5	-0.7	2.3	1.3	-0.3	0.7	-3.8	5.6	64.3
Hotan	-1	0.7	0.2	-0.1	0.1	-0.1	-0.4	0.9	-0.6	43.9	63.9	10.2
Kashi	-0.6	1.2	-1	-0.3	0	0	-0.9	1.2	-1	78.3	75.3	-16.5
Karamay	-0.4	0.1	1.4	-3.3	2	0.4	-1.3	0.1	0.1	72.9	92.9	-32.4
Artux	-0.6	2.1	-3.2	-2.3	0.4	1	-0.3	1.1	-0.8	81.7	29.2	22.9
Shihezi	-0.9	-0.3	2.2	-2.9	-1.5	3.8	-0.5	-0.2	0.4	-6.5	104.2	-41.8
Tacheng	-0.4	-0.2	1.4	-2.8	1.3	0.7	-0.3	-0.2	0.4	25.3	91.8	-38
Turpan	-1.2	0.3	1.8	-0.2	-0.9	1.1	-1.5	0.3	0.2	21.5	74.3	15
Urumqi	-1.2	-0.3	1.6	-2.6	-2	3.8	0.7	-0.9	1.2	-75	74.8	-7.9
Yining	-0.8	-0.2	2.7	-4	2.4	1.1	2	0.1	0.3	-83	58.3	37.7

[a] T1, T2, and T3 represent the years 2000-2005, 2005-2010, and 2010-2015.

Results show that EHT, AAT, and AP all conform to a fluctuating ascending trend in most areas of Xinjiang, while the ELT declined. On the whole, climate change is relatively mild while there has been a significant change in precipitation as shown in figure 1. Table 1 shows changes of climate in each region of Xinjiang. From 2000 to 2015, changes of temperature and precipitation exceeded 1°C and 50 mm in many regions, among them, climate change in *Karamay*, *Kashgar* and *Urumchi* are most obvious. Distribution of EHT has gradually increased from northwest to southwest; meanwhile, the distributions of ELT and AAT are characterized by a gradual decrease from south to north. Of these, EHT in the *Turpan* was the highest, ELT and AAT in the Altay were lowest, and AAT in *Hotan* was the highest. In addition, because water vapor in this region comes mainly from the Atlantic and Arctic oceans, the precipitation distribution is characterized by enhancement from south to north and from east to west as shown in figure 2.

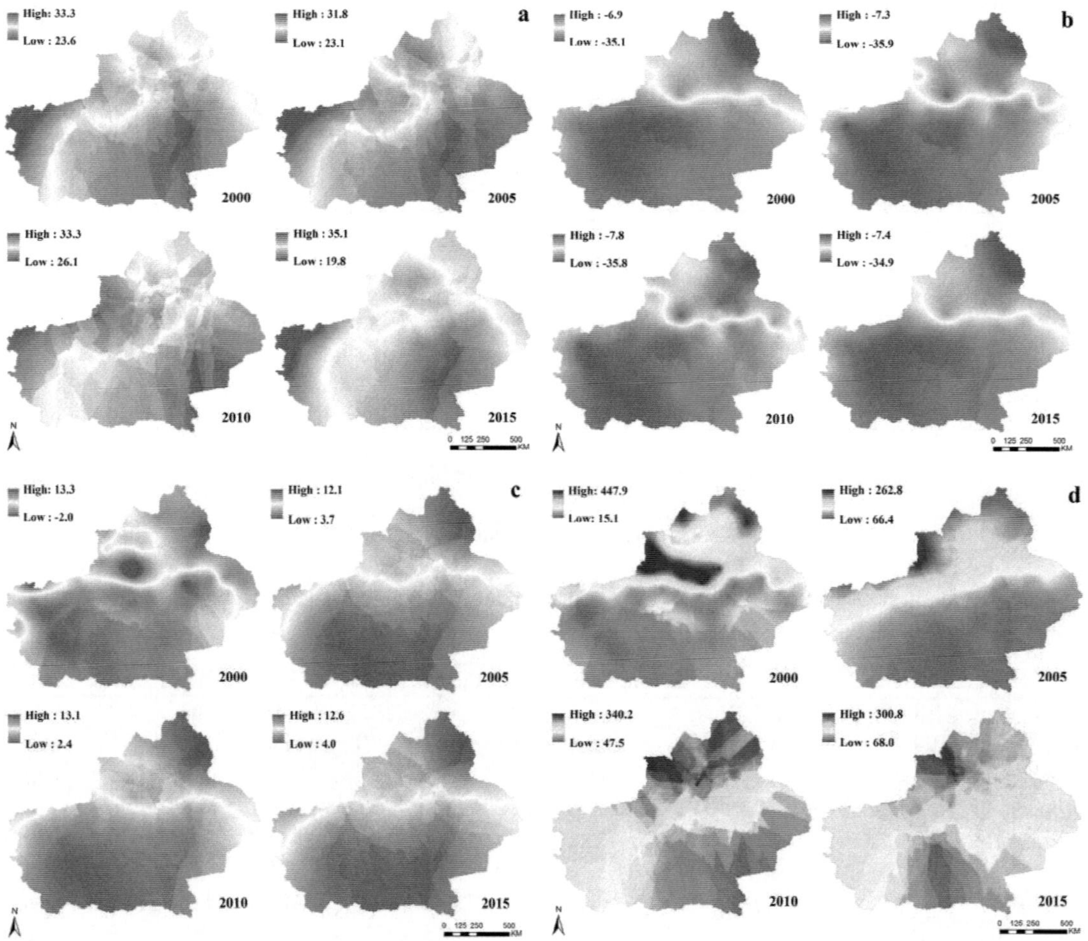

Figure 2. EHT (°C) (a), ELT (°C) (b), AAT (°C) (c), and AP (mm) (d) for Xinjiang (2000-2015).

3.2. Changes in agricultural landscape patterns

During 2000-2015, the pattern of agricultural landscape in Xinjiang in terms of area can be characterized as follows: grassland > farmland > water > forested land. Because of the impact of various natural, economic factors (GDP, industrial structure, etc.) and social factors (population, national policy, etc.), unused land in Xinjiang has been vigorously developed, resulting in a gradual increase in farmland area. In addition, the area of grassland has significantly declined, while areas of forested land and water first increased and then decreased over the time period of interest as shown in figure 3. Overall, the total area considered as agricultural landscape has tended to grow as shown in table 2, with the major source being development and utilization of unused land. As a result of the continuous development of western regions, agricultural development in Xinjiang has gradually expanded.

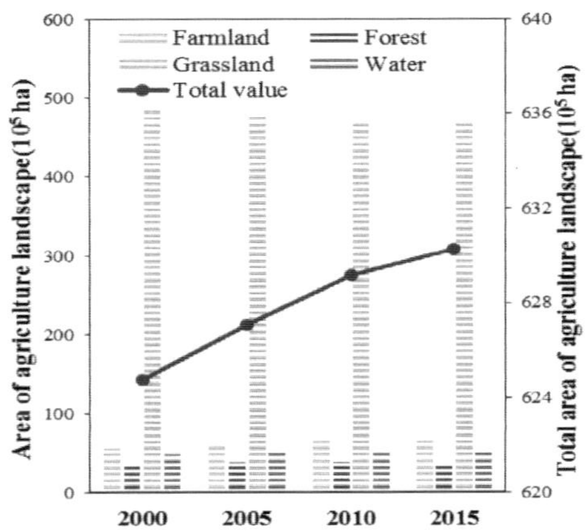

Figure 3. Land use in Xinjiang (2000-2015).

Table 2. Agriculture landscape metrics in Xinjiang in the period between 2000 and 2015.

	2000	2005	2010	2015	Change rate (%)		
					2000-2005	2005-2010	2010-2015
TA (1000 ha)	62475.3	62707.8	62918.3	63025.7	0.4	0.3	0.2
PD	0.014	0.015	0.016	0.016	13.2	1.9	1.3
CONTAG (%)	60.5	59.5	59.1	58.9	-1.6	-0.7	-0.3
SHDI	0.97	0.98	1	1.01	1.2	1.1	0.3
AI (%)	88.2	87.6	87.5	87.4	-0.7	-0.1	-0.1

Results show that both PD and SHDI have gradually increased as shown in figure 4 and table 2. Landscape fragmentation and heterogeneity within agricultural landscape patches has increased, and this landscape type has become more abundant. A gradual reduction in CONTAG has resulted in connectivity of landscape mosaics as well as a balanced spatial distribution of various landscape patches. In addition, a gradual decrease in AI has translated into a reduction in aggregation degree of the same patch type. During 2000-2015, the change rates of the various landscape metrics have continuously lowered, and there has been a more substantial landscape pattern change in northern Xinjiang compared to the south. This result shows that the agricultural landscape pattern in Xinjiang has gradually stabilized, while development in the northern regions is faster than in the south.

3.3. Changes in ecosystem service values

As a result of the development of eco-agriculture in Xinjiang, development efforts in the inland desert area have been re-invigorated, leading to constant increases in both farmland area and crop yield. In addition, substantial increase in prices has led to an increase in the value of an agricultural ecological unit equivalent factor in the period 2000 to 2015, 1034 Yuan/hm^2, 1060 Yuan/hm^2, 1713 Yuan/hm^2, and 2159 Yuan/hm^2, calculated using equation (2) for the years 2000, 2005, 2010, and 2015, respectively.

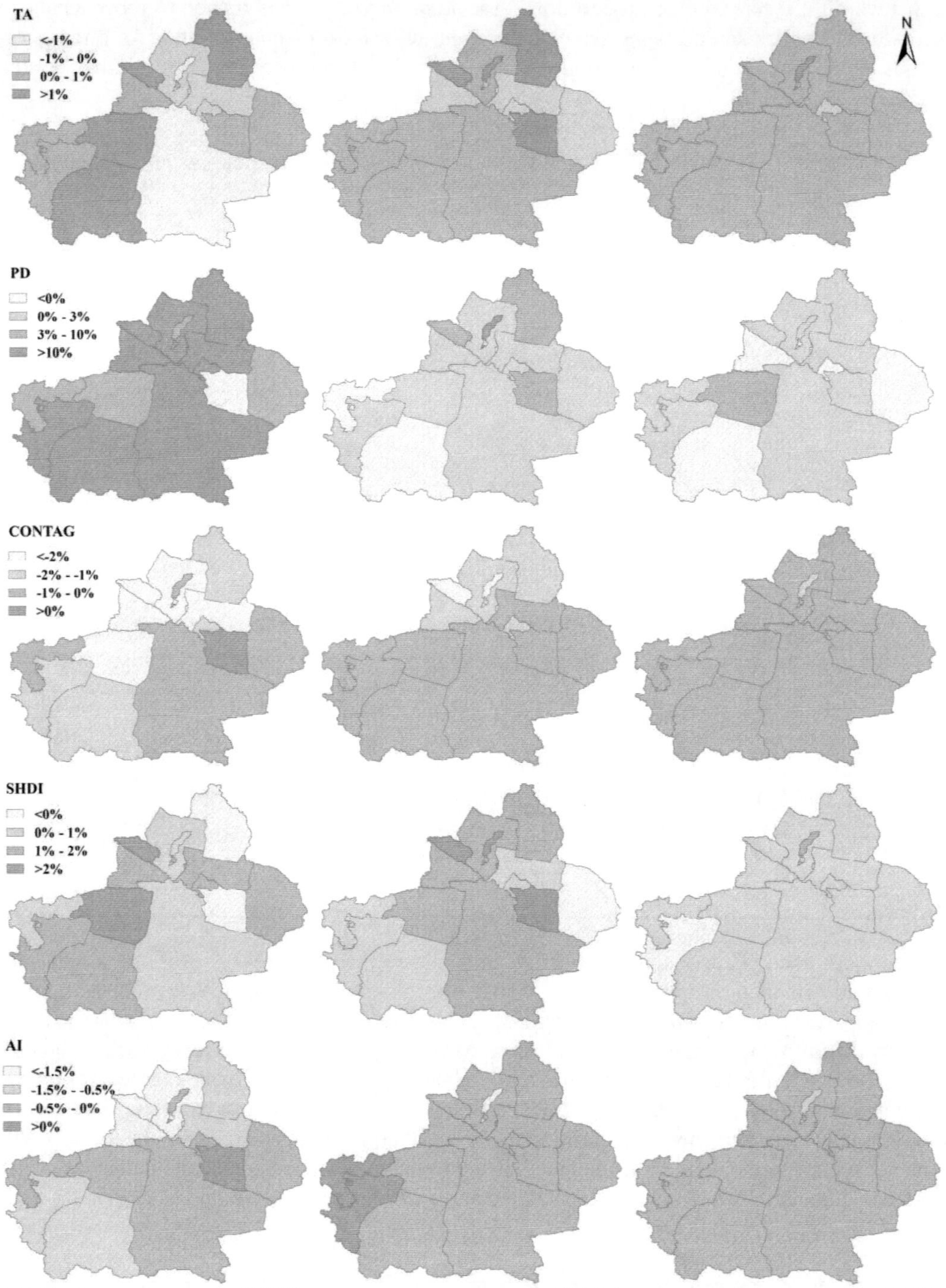

Figure 4. Changes of agricultural landscape metrics in each region during different temporal intervals.

Results indicate that the total AESV for this region has continuously improved, from 715.4 billion RMB Yuan to 1528.7 billion RMB Yuan, in the period 2000 to 2015. In terms of different land use types, the ecosystem service value for grassland is the highest, followed by water, forested land and farmland as shown in figure 5(a). Ecosystem service values for water, forested land, and farmland have continuously increased, while this value for grassland declined in 2010. In sum, ecosystem service values produced on the basis of ecological function can be summarized as follows: waste disposal > water retention > soil formation and protection > biodiversity conservation > climate regulation > gas regulation > entertainment and culture > food production > raw materials as shown in figure 5(b), reflecting the large waste disposal capacity of forests and grassland, and the regulatory function of water bodies in Xinjiang.

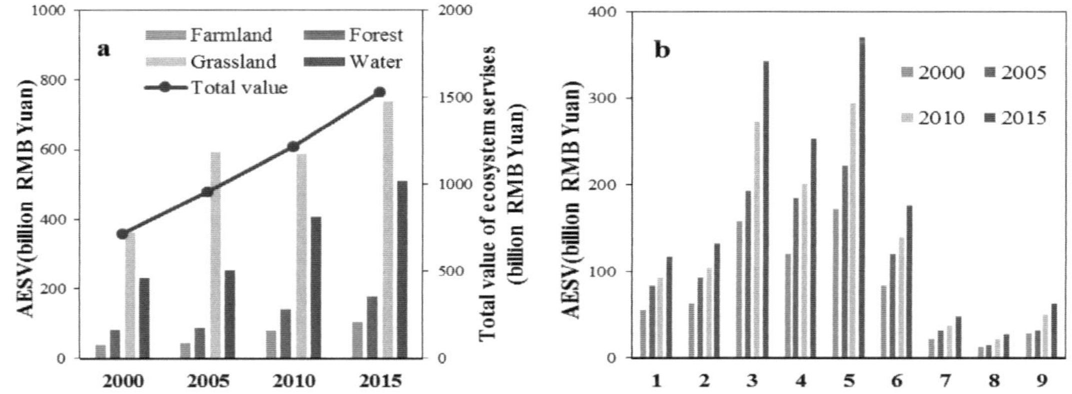

Figure 5. AESV during 2000-2015.

The numbers 1-9 refer to gas regulation (1), climate regulation (2), water reservation (3), soil formation and protection (4), waste disposal (5), biodiversity conservation (6), food production (7), raw materials (8), and entertainment and culture (9).

3.4. Agricultural landscape patterns in response to climate change

Using the agricultural landscape metrics and meteorological data for the 15 prefectures in Xinjiang (listed in table 1) for the years 2000, 2005, 2010, and 2015, the relationships between meteorological factors and agricultural landscape patterns were quantitatively analyzed using Pearson's correlation coefficients. Generally, EHT, ELT and AAT were significantly positively correlated to TA, CONTAG, and AI, and have a significant negative correlation to PD and SHDI. Of these, the highest correlation was observed between average temperature and TA, while the rise in average temperature extends into the allowable range of agricultural development, resulting in an increase agricultural landscape area. In contrast, AP was significantly negatively correlated with TA, CONTAG, and AI, but showed an obviously positive correlation to PD and SHDI. During 2000-2015, a changing pattern in AP is most significantly correlated with CONTAG, SHDI, and AI as shown in table 3.

Results show that landscape diversity, the extent of landscape fragmentation, and heterogeneity within patches decreased while aggregation increased as temperature rose. In contrast, an increase in precipitation led to increases in landscape diversity and fragmentation, as well as a reduction in heterogeneity within patches and aggregation. In the period between 2000 and 2015, both temperature and precipitation increased in Xinjiang while PD and SHDI rose, and CONTAG and AI decreased. These results suggest an enhanced effect of precipitation on the agricultural landscape patterns in the area.

Table 3. Pearson's correlation coefficients between landscape metrics and climate characteristics.

	TA	PD	CONTAG	SHDI	AI
EHT	ns[a]	ns	0.40**	-0.39**	0.41**
ELT	0.49** [b]	-0.42**	0.39**	-0.40**	0.36**
AAT	0.53**	-0.39**	0.456**	-0.47**	0.43**
AP	-0.52**	-0.47**	-0.63**	0.62**	-0.61**

[a]No significant correlation; [b]** p<0.01.

4. Conclusion

This study analyzed climate trends and spatiotemporal changes in agricultural landscape patterns in Xinjiang over the last 15 years, and evaluated the AESV. In addition, correlation analyses were used to quantify the association between agricultural landscape patterns and climate change. The study provides the scientific basis for the development of ecological agriculture, and promotes the sustainable development of agriculture into the future. However, because landscape pattern variations are the manifestation of the mutual interaction of natural environmental and human activities, the future work is to quantitatively analyze the driving force and influence mechanism.

Acknowledgments
The study was supported as a data compilation and processing project, National Special Program on Basic Works for Science and Technology of China (No. 2013FY110900).

References

[1] McGinlay J, Gowing D J G and Budds J 2017 The threat of abandonment in socio-ecological landscapes: Farmers' motivations and perspectives on high nature value grassland conservation *Environ. Sci. Polic.* **69** 39–49
[2] Rodrigo C J, Iserloh T and Lassu T 2016 Quantitative comparison of initial soil erosion processes and runoff generation in Spanish and German vineyards *Sci. Total Environ.* **565** 1165–74
[3] Tieskens K F, Schulp C J E and Levers C 2017 Characterizing European cultural landscapes: Accounting for structure, management intensity and value of agricultural and forest landscapes *Land Use Polic.* **62** 29–39
[4] Hietel E, Waldhardt R and Otte A 2004 Analysing land-cover changes in relation to environmental variables in Hesse, Germany *Landscape Ecol.* **19** 473–89
[5] Bürgi M, Hersperger A M and Schneeberger N 2004 Driving forces of landscape change-current and new directions *Landscape Ecol.* **19** 857–68
[6] Qadir M, Noble A D and Chartres C 2013 Adapting to climate change by improving water productivity of soils in dry areas *Land Degrad. Dev.* **24** 12–21
[7] Guo J 2015 Advances in impacts of climate change on agricultural production in China *J. Appl. Meteor. Sci.* **26** 1–11
[8] Liu Y S, Liu Y and Guo L Y 2010 Impact of climatic change on agricultural production and response strategies in China *Chin. J. Eco-Agricul.* **18** 905–10
[9] Xiao G J, Zhang Q and Wang J 2007 Impact of global climate change on agro-ecosystem *Chin. J. Appl. Ecol.* **18** 1877
[10] Keesstra S D, Geissen V, Mosse K, Piiranen S, Scudiero E and Leistra M 2012 Soil as a filter for groundwater quality *Curr. Opin. Env. Sust.* **4** 507–16
[11] Costanza R D, Arge R and Groot R D 1989 The value of the world's ecosystem services and natural capital *Ecol. Econ.* **387** 253–60
[12] Xu C, Chen Y, Yang Y, Hao X and Shen Y 2010 Hydrology and water resources variation and its response to regional climate change in Xinjiang *J. Geogr. Sci.* **20** 599–612
[13] Huang D L 2016 "Silk road" economic belt construction impact on China's agricultural economy in Xinjiang *World Agri.* **11** 31–9
[14] Su S J 2012 Ecological agriculture development present situation and countermeasure research in Xinjiang *J. Xinxiang Vocational Uni.* **20** 26–31
[15] Peng S, Du Q, Lin A, Hu B, Xiao K and Xi Y 2015 Observation and estimation of photosynthetically active radiation in Lhasa (Tibetan Plateau) *Adv. Space Res.* **55** 1604–12
[16] Xie G D, Lu C X and Leng Y F 2003 Ecological Assets Valuation of the Tibetan Plateau *J. Natur. Resour.* **18** 189–96

Characteristics of soil stability and carbon sequestration under water storage and drainage model

J Li[1, 2, 3], J C Han[1, 3, 4], C Chen[1, 3] and J J Yang[1, 3]

[1] Shaanxi Province Land Engineering Construction Group, Xi'an 710075, China
[2] College of Forestry, Northwest A & F University, Yangling 712100, China
[3] Key Laboratory of Degraded and Unused Land Consolidation Engineering, The Ministry of Land and Resources of China, Xi'an 710075, China

E-mail: kele502514@163.com

Abstract. This research was conducted to investigate the influence of saline alkali soil on soil physical properties, stability and organic carbon storage under water storage and drainage, and to provide scientific basis for improving soil quality in Fuping County of Shaanxi Province, China. Saline alkali soil model test was conducted and the process was assessed with two different methods: i) traditional drainage and ecological water storage, measure and analyze 0-30 cm soil bulk density, porosity, field water capacity, mean mass diameter (MWD), geological mean diameter (GMD), stability of water stable aggregate (WASR), aggregate destruction rate (PAD), fractal dimension (D) and; ii) organic carbon storage, comprehensively analyze the relationship between stability index and soil organic carbon. The results show that: (1) compared with traditional drainage treatment, water treatment may effectively reduce the soil bulk density by 1.3%-4.2%, and improve soil porosity and field capacity at the same time; (2) under dry and wet screen treatment, soil stability, the water storing treatment is higher than the drainage treatment. Performance trend of soil MWD and GMD increases with the increase of soil depth. The stability of soil water stable aggregates increased 14.5%-53.4%. The average aggregate destruction rate was 3.2% lower than that of the drainage treatment and the difference is obvious ($P< 0.05$). (3) Soil organic carbon content and organic carbon storage in 0-30 cm soil layer could be increased effectively by water storage. Both of them were 13.4%-27.9% and 9.9%-18.8% higher than the drainage treatment. (4) There is a negative correlation among average aggregate destruction rate, fractal dimension and soil organic carbon storage. The correlation coefficient is, respectively, R^2=0.86 and R^2=0.94, and the difference is obvious ($P<0.05$). To sum up, the water storage treatment can effectively improve the soil quality, improve soil stability and soil organic carbon storage, which can be a good control of saline alkali soil.

1. Introduction

In agricultural production, saline soil is a low yield soil, which is an obstacle to the construction of high and stable yield fields [1-3]. In particular, the threat of soil secondary salinization is highly extended in modern irrigated areas because of human activities [4-6]. Soil quality degradation may further cause groundwater pollution [7, 8], posing high health risks to residents and ecological diversity [9-11]. The comprehensive treatment of saline alkali land includes many methods, such as physical improvement, water conservancy, chemical and biological improvement. Water improvement, namely the model of "Fresh water salt and Drainage of salt", is mainly used in China. The practice of "drainage" increases the slat in the area. The traditional saline alkali land treatment technology incurs high operating cost, but also causes a waste of water resources and pollutes the downstream water to some extent [12-16].

The evaluation of soil quality generally includes a comprehensive analysis of soil bulk density, porosity and stability and carbon storage [17]. Soil structure controls soil stability. Aggregate is the basic unit of soil structure [18, 19]. The basic characteristic of soil is one of the key indicators determining the process and effect of soil separation, handling, compaction and compaction in soil eroding process, and an important evaluation index of soil safety, soil quality and health [20]. The stability of soil aggregates can be divided into mechanical stability, water stability and biological stability [21]. Generally, the stability index is used to characterize the soil structure. And it is related to the movement of water and the growth of crop root and the carrier of organics [22]. For the present stage, under the development of the saline alkali soil, different treatment methods affect the soil aggregation stability and water stability of soil aggregate distribution, and then affect the improvement and restoration of land quality. Soil aggregate stability is usually characterized by average diameter of soil water stable aggregates, geometric mean diameter and percentage of soil aggregates with diameter > 0.25 mm. Greater value indicates better aggregate stability [23]. Several researches mainly focused on the different cultivation patterns on soil aggregate and the impact on the carbon and land consolidation after a variety of land use mode studying the distribution characteristics of soil aggregate stability [24-28], but for a different approach to governance and saline-alkali soil on the stability of few reports research remains to be further strengthened.

Halogen Beach between Fuping County and Pucheng County in Shaanxi Province, China is affected by severe saline alkali soil. Lubotan soil, a typical soil type in the Halogen Beach, is mainly composed of moderately and severely salinized soil. Han et al. who proposed the traditional Saline alkali land treatment give priority to the water storage and establishing and perfecting irrigation and drainage system [29], storage fresh water on saline alkali land for leaching, washing salinity by irrigation, draining the saline water. They proposed a new model of "changing drainage to storage, water and land harmony, and harmonious ecology" for Saline alkali land treatment under water storage and drainage conditions. The research will set up simulation test in the Fuping experimental base of Shaanxi province, China to analyze the changes in soil physical characteristics and stability and organic content under different treatment modes. Comprehensive system research will be conducted on each treatment and conduct correlation analysis between the soil stability indicator and organic carbon, and provide scientific basis for the economic development and utilization of saline-alkali soil and improving the stability.

2. Materials and methods

2.1. Site description

The test is set up in Fuping experimental base (109°11′N, 34°42′E, 492 m above sea level), which is located in Chuyuan Village, Ducun Town, Fuping County, Weinan City, Shaanxi Province. This area belongs to the warm temperate semi humid climate zone. Annual rainfall is 473 mm on average and the rainfall in July-September covers 49% of the rainfall throughout the year. Annual evaporation is 1 000~1 300 mm and the frost free period is 225 d. Annual mean temperature is 13.4℃. The highest temperature in summer is 41.8℃ and the lowest temperature in winter is -22℃. The total amount of annual light radiation is 123.9～127.8 kca cm^{-2}.

2.2. Experimental design

The experiment was conducted in May 2010. In order to simulate the land condition in the experimental area of Halogen Beach, the test device is customized and the size is 23.0 m×1.5 m×2.0 m (length× width× height). It is made of cement brick and concrete structure. The test device includes two parts, namely water storage (water drainage) and test soil. One end stores water with water tank (drain water) and the middle section is soil tank. The water and soil are blocked with Geotextile to prevent the inlet of soil into the water surface. In order to ensure the test results, the interface of each device will conduct waterproof treatment. The experimental treatment is designed for two kinds of water storage and drainage. Water treatment is to put quantitative water in the ditch and take use of the dominant effect of "water- soil" dispersion effect to control the shift of soil salt; the drainage treatment takes use of broad irrigation and convection effect between water-soil to drain water and leach salt. The water quantity used in the water storage and drainage is the same. The salt shifting and changing under each treatment are shown in figure 1.

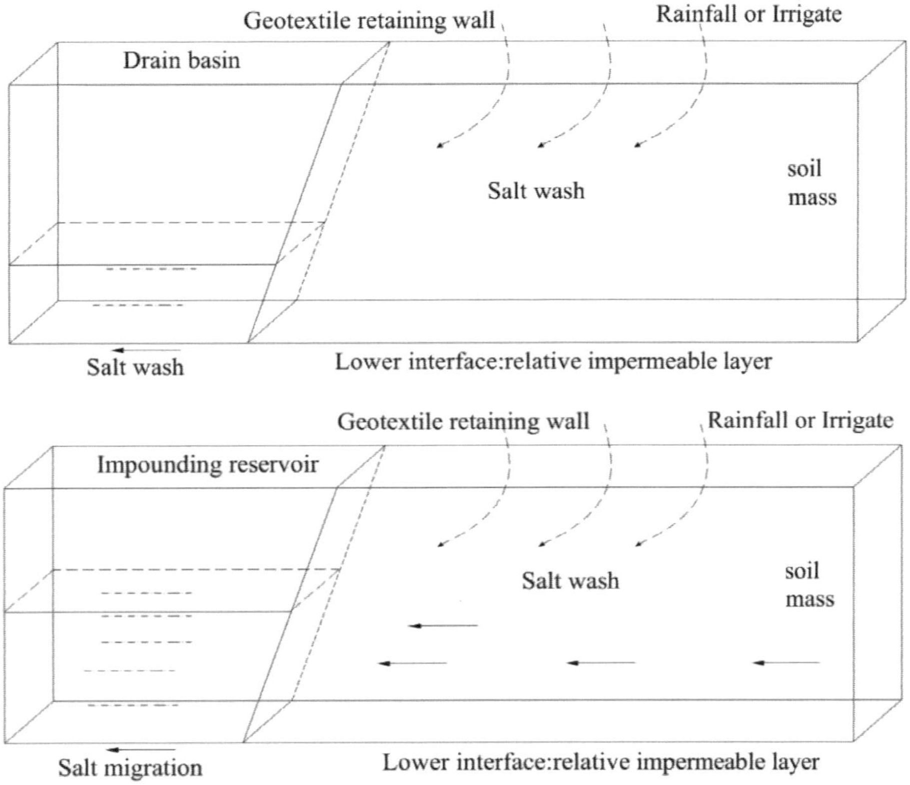

Figure 1. The soil salinity transport under water storage and water drainage treatments.

The test soil samples were selected from saline alkali soil in Halogen Beach. Halogen beach undisturbed soil was taken every 30-40 cm. the soil passes rolling, crushing, drying, sieving and fills in self-made test soil trough. According to the detection data of development soil in test zone of Northwest Agriculture and Forestry University Testing Center in 2010, soil analysis showed: Average content of soil organic matter is 0.7%; Total salt content is 0.8% on average; pH value is 9.3 on average; average content of Cl^-, HCO_3^- and SO_4^{2-} are 0.4%, 0.1% and 0.1%, respectively.

2.3. Measurement items and methods

2.3.1. Determination of physical index of soil. After the winter wheat is harvest, take the standard ring shear with height 5cm and diameter of 5.04 cm to collect the soil sample in each test plot for the determination of soil capacity and porosity. According to the diagonal sampling method, 4 sampling points were set up. The sampling depth and interval are respectively 30 cm and 10 m. Each plot takes a total of 24 bulk density samples. The average value determined with drying method refers to the soil capacity of each soil layer.

After the spring corn is harvest, 4 sampling points are set up according to the diagonal sampling method. The sampling depth and interval are respectively 30 cm and 10 cm. The undisturbed sample collected from each soil layer is used to determine relevant index of soil aggregate. There is not backlog and damage back to the laboratory. After natural drying, the stone and sundries existing in soil sample are removed. Dry sieving method is used to measure the mechanical stability of soil and wet sieving method is used to determine the water stability of soil [30].

2.3.2. Determination of chemical index of soil. Soil samples were collected immediately after winter wheat harvest. The soil depth of each plot was 0-30 cm and the sampling interval is 10 cm. the diagonal line method is used and 4 sampling points are set. Four repeated samples at the sample depth form one mixed sample and are naturally dried in the laboratory to remove gravel and plant roots, stubble and other debris and pass through 0.25 mm sieve. Potassium dichromate ($K_2Cr_2O_7$) by oxidation external heating method (GB7857-1997) is used to determine the soil organic carbon and calculate the reserve of organic carbon [31].

2.4. Data calculation

Soil bulk density, porosity, field capacity can be calculated as follows:

$$P=(DS-W)/V \tag{1}$$

Where P is soil bulk density (g/cm^3); DS is total weight of drying soil and cutting ring (g); W is the quality of cutting ring (g); V is the volume of cutting ring (cm^3)

$$SD=(1-P/SP)\times 100 \tag{2}$$

Where SD is the soil porosity (%); SP is specific gravity of soil (2.65 g/cm^3)

$$W_S=(M_d- DS)/(DS-W) \times 100 \tag{3}$$

Where W_S is field capacity (%) ; M_d is total weight of soil and ring cutter after water control (g)

To calculate the soil aggregate stability index, the equations derived by Yang *et al* were used [32].

$$Wn=\frac{W_{dn}/W_{wn}}{200}\times 100\% \tag{4}$$

$$R_{0.25}=\sum_{n=1}^{i}(W_n) \tag{5}$$

$$MWD=\sum_{n=1}^{i}(\overline{X_n}\times W_n) \tag{6}$$

$$GMD=\sum_{n=1}^{i}(\overline{X_n}\times W_n) \tag{7}$$

$$WSAR=WSA/MSA\times 100\% \tag{8}$$

$$PAD=(DR_{0.25}-WR_{0.25})/DR_{0.25}\times 100\% \tag{9}$$

$$\frac{M(r<\overline{X_{n)}}}{M_t}=(\frac{\overline{X_n}}{X_{\max}})^{3-D} \tag{10}$$

Where $R_{0.25}$ is the content of aggregate with diameter > 0. 25 mm (%); WMD is mean weight diameter of aggregate (mm); GMD is the graphical mean diameter of aggregate (mm); X_n is the mean diameter of aggregate in certain scope; *Wn* is the proportion of n level aggregate mass (%); dry sieving (W_{dn}) and wet sieving method (W_{dn}) may calculate them. $M(r< X_n)$ is the weight of aggregate with grain size less than X_n (g); *Mt* is total weight of aggregate (g); *Xmax* is the maximum particle size of aggregate (mm); D is aggregate fractal dimension.

Organic carbon storage is calculated with the following equation by Liang *et al* [26].

$$SOC_{(storage\ in\ depth)}=2\times P\times C \tag{11}$$

Where: $SOC_{(storage\ in\ depth)}$ is the unit area of soil carbon reserve in equal depth (t hm^{-2}); *P* is the capacity of soil (t/hm^2); C is the content of soil organic in 0-10, 10-20 and 20-30 cm layer (g kg^{-1}).

2.5. Data analysis

The test data were analyzed by using SPSS (PASW Statistics 16.0) and factor analysis of variance were performed. EXCEL 2007 and Sigmaplo12.5 were used to organize and map these data.

3. Results and analysis

3.1. Analysis on soil physical characteristics of water storage and drainage mode

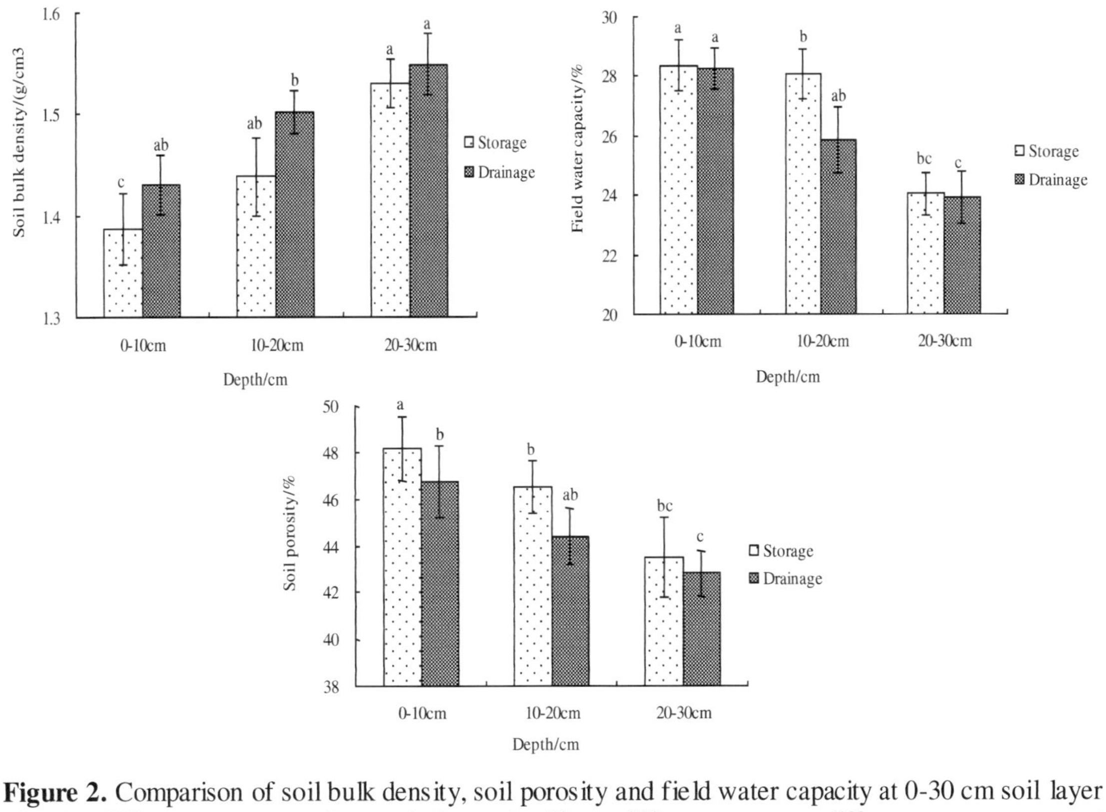

Figure 2. Comparison of soil bulk density, soil porosity and field water capacity at 0-30 cm soil layer of under different treatments. Letters indicate significant differences among different treatments at $p < 0.05$ levels.

Different treatments had significant effects on soil bulk density, soil water holding capacity and soil porosity (figure 2). Soil water holding capacity and porosity were negatively correlated with soil bulk density. If the unit weight is larger, the soil will be denser, thus field capacity and porosity will be decreased. In 0-30cm soil layer, the soil bulk density increased with the increase of soil depth. There is significant difference between water storage and drainage, so is the soil layers (*P<0.05*). In 0-30 cm soil layer, the soil capacity of water storage treatment is 1.3%-4.2% lower than the drainage treatment. As for the field capacity, the water storage and drainage treatment perform consistently, and decrease with the increase of soil depth. The field capacity of water storage treatment is superior to the water drainage treatment. The average field capacity under storage treatment is 3.1% higher than the water drainage treatment. The trend of soil porosity under different treatments was consistent with field capacity and decreases with the increase of soil layer. The water storage treatment is 3.1%, 4.7% and 1.5% higher than drainage treatment in soil porosity. Upon the comprehensive analysis of soil bulk density, field capacity and soil porosity under two treatment modes, water storage can effectively improve the soil porosity, increase soil aeration and water permeability, and is beneficial to maintain soil bulk density at a low level; drainage treatment could increase the soil bulk density and promote the soil hardening.

3.2. Analysis on characteristics of soil aggregates in water storage and drainage model

3.2.1. Analysis of soil aggregate size. From the perspective of WMD, WMD and GMD under dry sieving of each treatment method are far higher than WMD and GMD under wet sieving (table 1) mainly because most of the soil aggregates in the experiment were non water stable aggregates. The performance tendency of WMD and GMD of soil under dry sieving method and wet sieving method is similar. The water storage treatment is higher than the drainage treatment and MWD and GMD of soil increases with the soil depth.

Table 1. MWD and GMD of dry and wet sieving under different pattern.

Method	Index	Treatment	Soil layer (cm)		
			0-10	10-20	20-30
Wet sieving	WMD	Storage	5.65±0.25ab	5.50±0.03ab	6.38±0.23a
		Drainage	4.80±1.03b	5.04±0.23ab	5.90±0.24ab
	GMD	Storage	4.07±0.37abc	4.01±0.04abc	5.35±0.36a
		Drainage	2.77±0.81c	3.24±0.29bc	4.50±0.30ab
Water sieving	WMD	Storage	0.217±0.01b	0.224±0.02b	0.222±0.01a
		Drainage	0.184±0.03b	0.207±0.12b	0.216±0.04b
	GMD	Storage	0.140±0.00b	0.154±0.01ab	0.182±0.01a
		Drainage	0.139±0.00b	0.148±0.02b	0.153±0.01ab

Note: MWD: mean weight diameter, GWD: mean geometric diameter.
values followed by the different letters are significantly different at *P <0.05*, the data: mean±standard deviation. The same below.

The difference of WMD and GMD of two treatment methods under dry sieving is obvious (*P< 0.05*). The average weight diameter of soil layers under water storage treatment was higher than that under drainage treatment. In 0-30 cm soil layer under water storage treatment, WMD increases by 8.1%-17.7% than drainage treatment; the increase amplitude of GMD is 18.9%-46.6%. The difference of two treatment methods under the wet sieving treatment is not obvious as the dry sieving treatment. The effect of water storage treatment in improving soil structure is superior to the drainage treatment. Under wet sieving, WMD is 17.8% higher than the drainage treatment in 0-10 cm soil layer under water storage treatment. The improvement effect is the most obvious compared with other soils. It is mainly because the salt is leached after broad irrigation, thus causing serious compaction on soil surface. The soil moisture can be effectively increased by the water storage treatment, and the soil water stability is effectively improved. In 0-10 cm and 10-20 cm soil layer, the water storage treatment increases GMD by 1.1% and 4.3% than the drainage treatment. There is significant difference between two treatment methods (*P<0.05*). WMD and GMD are the common indexes reflecting the size and distribution of soil aggregate.

3.2.2. Soil aggregate stability analysis. Under the treatment methods, the trend of WSAR and PAD is shown in figure 3. WSAR decreases with the increase of soil depth and PAD increases with the increase of soil depth. In 0-10, 10-20 and 20-30 cm soil layer, WSAR under water storage treatment is 30.2%, 53.4% and 14.5% higher than that under water drainage treatment; PAD decreases by 30.2%, 53.4% and 14.5% respectively. Upon comprehensive analysis WSAR under water storage treatment is 33.9% higher than that in water drainage treatment, and PAD is 3.2% lower than the water drainage treatment. The difference between two treatment methods is obvious (*P<0.05*). Compared with water drainage treatment, water storage treatment may effectively improve soil stability, and has a positive effect on increasing soil aggregate structure and improving inter aggregate cementation.

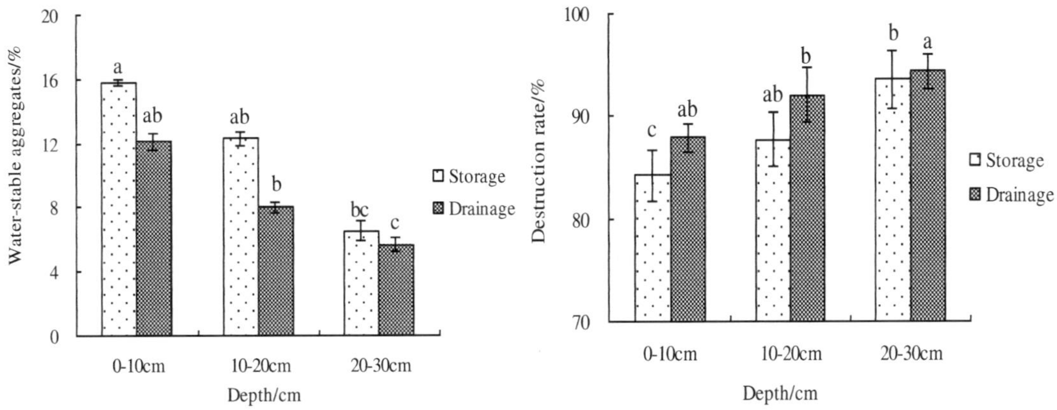

Figure 3. WSAR and destruction rate of aggregate (%) under different pattern.

Table 2. Fractal dimension of soil aggregates under different pattern at 0-30cm soil layer.

Treatment	Soil layer (cm)	Fractal dimension (wet sieving)	Fractal dimension (water sieving)
Storage	0-10	1.97±0.01ab	2.93±0.21c
	10-20	1.90±0.03ab	2.95±0.11bc
	20-30	1.65±0.06b	2.96±0.05ab
Drainage	0-10	2.15±0.09a	2.94±0.07bc
	10-20	2.04±0.05a	2.97±0.04b
	20-30	1.84±0.04ab	2.98±0.01a

3.2.3. Fractal dimension analysis. The fractal dimension of soil is the parameter reflecting the geometry of soil structure. The fractal dimension of soil aggregate size distribution reflects the effect of soil water stable aggregates and water stable aggregates on soil structure and stability, namely the smaller the fractal dimension, the better the structure and stability of soil [33]. Equation (10) is used to conduct fitting calculation on the dry sieving and wet sieving data of aggregate to obtain fractal dimension value. The fractural dimension of aggregate of sectional soil in 0-40 cm soil under different treatment methods is shown in table 2. In 0-30 cm soil layer, the fractal dimension of soil aggregate of each treatment in dry sieving method decreases with the increase in soil depth and the performance tendency is consistent. The fractal dimension of soil aggregate is 1.65-2.15 and the difference of two treatment methods is obvious (*P<0.05*). In the soil stability, the water storage treatment is 7.0%-11.8% higher than the water drainage treatment; in 0-30 cm soil, the average stability under water storage treatment is increased by 9.1% than water drainage treatment. In 0-30 cm soil layer, the fractural dimension of soil aggregate with wet sieving method is opposite to the dry sieving method, namely each treatment increases with the increase of soil depth. The fractural dimension of soil aggregate is 2.93-2.98.

3.3. Effects of organic carbon and organic carbon storage

The difference of organic carbon and organic carbon storage of different layers of soil under water storage and drainage treatment methods is shown in table 3. The tendency of organic carbon and organic carbon storage is the same, and show negative relationship with the soil depth. It decreases with the increase of soil depth. The organic carbon and organic carbon storage in water storage treatment are higher than the water drainage treatment. In 0-30 cm soil layer, the organic carbon content under water storage treatment is 13.4%-27.9% higher than the water drainage treatment; the organic carbon storage is 9.9%-18.8% higher than the water drainage treatment. In 0-30 cm soil layer, organic carbon and organic carbon storage under water storage treatment are 18.9% and 15.8% higher than the water drainage treatment and the difference is obvious (*P<0.05*). The research results show that on the premise of other treatment measures are consistent, water storage and drainage treatment is helpful to improve the soil organic carbon content and carbon storage, and has a positive effect on the stability of soil.

Table 3. Soil organic carbon soil organic carbon storage under different pattern at 0-30 cm soil layer.

Treatments	Soil layer (cm)	Soil organic carbon (g kg^{-1})	Soil organic carbon storage (t hm^{-2})
Storage	0-10	6.08±0.35a	16.87±0.50a
	10-20	5.05±0.36bc	15.45±0.67a
	20-30	4.90±0.50bc	14.10±0.61a
Drainage	0-10	5.36±0.05ab	15.33±0.40a
	10-20	4.28±0.30bc	12.86±0.82b
	20-30	3.83±0.20c	11.87±0.14b

3.4. Relationship between organic carbon storage and soil stability index

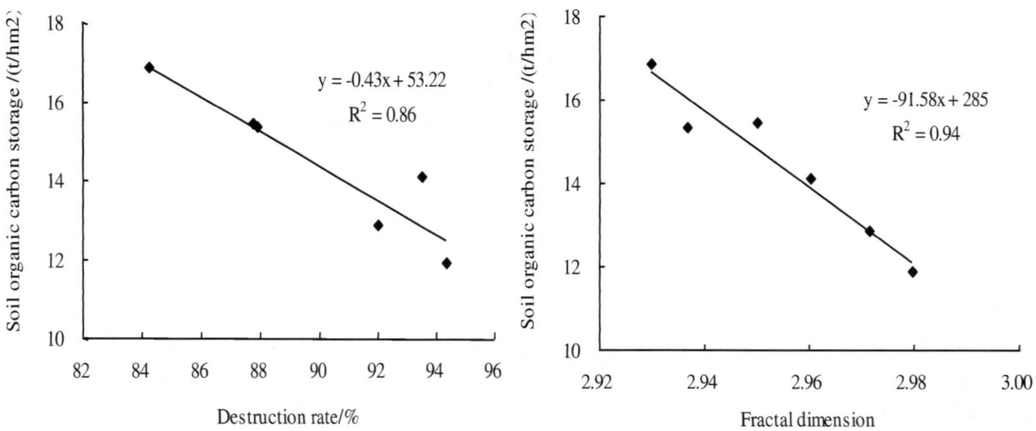

Figure 4. Comparison of soil fractal dimension and soil PAD and soil organic carbon storage under different treatments.

The relationship between average organic carbon and average fractal dimension in 0-30 cm soil layer in water storage and drainage treatment is shown in figure 4. There is negative relationship between organic carbon reserve and fractal dimension, namely when the fractal dimension is larger, the reserve of organic carbon will be smaller and it is not conducive to the improvement of soil stability. Under water storage and drainage treatment, we conduct fitness analysis on the PAD (X) and soil organic carbon storage (Y) in 0-30 cm layer. The regression equation Y=-0.43x+53.22 (R^2=0.86) and the difference is obvious (*P<0.05*); the regression equation exists between x and Y, namely Y=-91.58x+285 (R^2=0.94). The research results indicate organic carbon reserve, fractal dimension and PAD are important indicators reflecting the soil stability. There is high correlation among three factors. It may be directly regarded as the scientific basis of soil stability evaluation.

4. Discussion

4.1. Influence on the physical characteristics of soil

Soil bulk density, field capacity and porosity are one of the important indicators evaluating the soil quality [34, 35]. At present, the research on saline alkali land mainly focuses on the selection of salt tolerant varieties, the choice of irrigation methods and the design of saline alkali land treatment project. After cyclic salt and its engineering application, the spatial distribution of salt in the vertical direction can be changed, and the pressure can be gradually reduced. However, there are few reports on the changes of soil physical characteristics [36, 37]. Compared with drainage treatment, water storage treatment can effectively reduce soil salinity, effectively maintain soil compaction and soil bulk density in a suitable range. Straw returning treatment has obvious effect on improving the soil structure [38]. Water treatment can effectively improve the soil bulk density, increase soil water and improve the soil pore distribution. Combined with straw returning, reasonable distribution of solid, gas and liquid accelerated the decomposition of straw, and improved soil structure to some extent. Under water drainage treatment, the irrigation increased soil compaction and structural damage, and it is easy to form soil runoff channel, increase nutrient loss and reduce water use. The results showed that the water treatment significantly improved the soil quality.

4.2. Influence on the aggregate characteristics of soil

Previous studies have shown that there is a significant positive correlation between the fractal dimension of soil and the content of soil aggregates [39]. The storage and drainage treatment combined with straw returning could affect the transformation and redistribution of small aggregates and large aggregates, and then affect the stability and erosion resistance of soil structure. Many scholars have shown that there is a significant negative correlation between the fractal dimension of soil and large aggregates, namely the fractal dimension decreases with the increase of the content of large aggregates [40, 41]. The research shows that under dry sieving and wet sieving, WMD and GMD under water storage treatment are superior to the drainage treatment, and increase with the increase of soil. WSAR and PAD may better reflect the soil structure stability. If WSAR value is higher, it indicates the soil structure will be more stable; if PAD value is higher, it indicates the soil structure is not stable and the degradation degree increases. In 0-30 cm soil layer, WSAR under water storage treatment is more obvious than the water drainage (*P<0.05*), and the increase amplitude is 32.7%; soil PAD of water storage treatment is obviously lower than the water drainage treatment. It is beneficial to the improvement of soil structure and soil degradation. D is a new comprehensive index which evaluates the soil structure distribution. The index may not only express the size and distribution of soil particle, but also reflect the evenness degree of soil texture. The value is higher, indicating the texture is thinner and the soil penetration ability is weaker; the value is smaller, indicating the soil structure is better and the through ability is better. The research indicates in 0-30 cm soil and with dry sieving method, the fractal distribution of soil aggregate of water storage and drainage treatment decreases with the increase of soil depth and the performance tendency is consistent. D of soil aggregate is 1.65-2.15; with wet sieving method, D of soil is 2.93-2.98. It is opposite to the results obtained with dry sieving method, namely it increases with the soil depth.

4.3. Influence on organic carbon and organic carbon storage of soil

Application of various nutrient elements and other elements in soil contributes to the formation of small aggregates and its stability. The balanced application of fertilizer and good water retention are beneficial to the growth of plants and the formation of aggregates. Studies have shown that the chemical alkaline solution has a destructive effect on the large aggregates, which can disperse the particles, which is not conducive to the maintenance and improvement of the stability of the aggregates [42-44]. This is the same as the results of this study. Water storage can effectively reduce the soil salt content, and has obvious effect on improving the stability of soil. Water retention is beneficial to the decomposition and transformation of organic matter such as straw, which is beneficial to the increase of organic carbon and organic carbon in soil [45]. Straw returning to the field, the addition of plant residues effectively improved the organic matter status of the aggregates, and played a positive role in promoting the formation and stability of aggregates. The results showed that soil organic carbon content and storage were higher than that of water drainage treatment. Six thinks that large aggregates are formed around fresh plant residues, and the quantity of micro-aggregate will increase after the relieving the new and old micro-aggregate which are enclosed by macro aggregate [46]. After No tillage coverage, Soil organic matter input increased, soil organic carbon content increased, more new large aggregates will increase, while the micro aggregates decreased accordingly. The results show that SOC is closely related to soil aggregates, that is, the higher the soil stability, the higher the soil organic carbon content, the larger the fractal dimension of soil aggregates. Correlation coefficient of soil organic carbon storage is R^2=0.94. The studies have shown that the correlation coefficient between soil aggregate destruction rate and organic matter content is 0.89. The higher the soil organic matter content, the lower the aggregate destruction rate. The results of this study are consistent with the results that the soil organic matter content is negatively correlated with the aggregate destruction rate, and the relevant coefficient is R^2=0.84.

5. Conclusions

This study investigated the characteristics of soil stability and carbon sequestration under water storage and drainage. The following conclusions can be summarized:

- Water storage treatment may effectively improve the soil capacity, porosity and field capacity in 0-30cm soil and the improvement effect is better than the water drainage treatment.
- Water storage treatment is superior to water drainage treatment in mean mass diameter and geological mean diameter in 0-30 cm soil. It tends to increase with the increase of soil and stability of water stable aggregate, aggregate destruction rate and fractal dimension better reflect the soil stability. WASR of water storage treatment is 32.7% higher than water drainage treatment. It may effectively improve the soil stability.
- Water storage treatment may effectively improve the organic carbon and organic carbon storage of soil layer. There is negative relationship among organic carbon storage, PAD and D. the coefficient of correlation is R^2=0.84 and R^2=0.94.

To sum up, the water treatment is conducive to improving the soil quality in saline-alkali soil and. The effect on the improvement in soil stability is superior to the water drainage treatment. The stability of the soil in the deep soil of saline alkali soil needs to be further studied.

Acknowledgments
This study was supported by The key scientific and technological innovation team project in shaanxi province of China (2016KCT-23). We are grateful to Mr Han Jichang for providing laboratory, so we could managing the field experiments and professional English editor Cao Yu and kind help with the language of this manuscript.

References

[1] Gorji T, Sertel E and Tanik A 2017 Monitoring soil salinity via remote sensing technology under data scarce conditions: A case study from Turkey *Ecol. Indic.* **74** 384-91

[2] Manandhar R and Odeh I 2014 Interrelationships of land use/cover change and topography with soil acidity and salinity as indicators of land degradation *Land* **3** 282-99

[3] Wu J, Li P, Qian H and Fang Y 2014 Assessment of soil salinization based on a low-cost method and its influencing factors in a semi-arid agricultural area, northwest China *Environ. Earth Sci.* **71** 3465-75

[4] Li P, Qian H, Howard K and Wu J, 2015 Building a new and sustainable "Silk Road economic belt" *Environ. Earth Sci.* **74** 7267-70

[5] Li P, Wu J and Qian H 2016 Regulation of secondary soil salinization in semi-arid regions: A simulation research in the Nanshantaizi area along the Silk Road, northwest China *Environ. Earth Sci.* **75** 1-12

[6] Yang Z 1997 Land degradation and its control strategies in China *China Environ. Sci.* **17** 108-12

[7] Li P, Tian R, Xue C and Wu J 2017 Progress, opportunities and key fields for groundwater quality research under the impacts of human activities in China with a special focus on western China *Environ. Sci. Poll. Res.*1-11

[8] Li P, Qian H and Zhou W 2017 Finding harmony between the environment and humanity: An introduction to the thematic issue of the Silk Road *Environ. Earth Sci.* **76** 105

[9] Wu J and Sun Z 2016 Evaluation of shallow groundwater contamination and associated human health risk in an alluvial plain impacted by agricultural and industrial activities, mid-west China *Expo. Health* **8** 311-29

[10] Li P, Qian H and Wu J 2014 Origin and assessment of groundwater pollution and associated health risk: A case study in an industrial park, northwest China *Environ. Geochem. Health* **36** 693-712

[11] Li P, Li X, Meng X, Li M and Zhang Y 2016 Appraising groundwater quality and health risks from contamination in a semiarid region of northwest China *Expo. Health* **8** 361-79

[12] Liu J, Luo Z, Jia Z, Fang S and Wang N 2005 Feasibility study of adopting controlled drainage in Yinnan irrigation district based on salt and water balance *T. Chin. Society Agr. Eng.* **21** 43-5

[13] Yue W, Yang J, Tong X and Gao H 2008 Transfer and balance of water and salt in irrigation district of arid region *J. Hyd. Eng.* **39** 623-8

[14] Yang G 1999 A study on water salt of Solonchak in Hetao plan, Inner Mongolia *Sci. Silvae Sinicae* **35** 107-10

[15] Lv D, Wang Q, Wang W and Shao M 2000 Soil water and salt transport features under one dimension infiltration *J. Soil Water Conserv.* **14** 91-4

[16] Wang Q and Wang W 2000 Water and salt transport features for salt effected soil through drip irrigation under film *Trans. Chin. Society Agr. Eng.* **16** 54-7

[17] Khaledian Y, Kiani F, Ebrahimi S, Brevik E and Aitkenhead-Peterson J 2016 Assessment and monitoring of soil degradation during land use change using multivariate analysis *Land Degrad. Dev.* **28** 128-41.

[18] Lado M, Paz A and Ben-Hur M 2004 Organic matter and aggregate-size interactions in saturated hydraulic conductivity *Soil Sci. Soc. Am. J.* **68** 234

[19] Taboada M, Barbosa O, Rodríguez M and Cosentino D 2004 Mechanisms of aggregation in a silty loam under different simulated management regimes *Geoderma* **123** 233-44

[20] Rodrigo Comino J, Ruiz Sinoga J, Senciales González J, Guerra-Merchán A, Seeger M and Ries J 2016 High variability of soil erosion and hydrological processes in Mediterranean hillslope vineyards (Montes de Málaga, Spain) *Catena* **145** 274-84

[21] Lu J and Li Z 2002 Advance in soil aggregate study *Res. Soil Water Conserv.* **9** 81-5

[22] Yuan X, Jiao J, Zhu L, Liu M, Li H and Hu F 2011 Effects of earthworm activity on soil aggregates' stability and organic carbon distribution under different manipulations of corn straw *Soils* **43** 968-74

[23] Wei Y, Ma T, Wei X, Wang C, Hao M and Zhang M 2016 Effects of cropping systems on distribution of water-stable aggregates and organic carbon and nitrogen in soils in semiarid farmland of the Loess Plateau *J. Agro-Environ. Sci.* **35** 305-13
[24] Sun G, Chen F, Li L, Wu F, Xiao X and Zhang H 2007 Effects of tillage on the carbon pool of paddy soil with long-term no-tillage *J. China Agr. U.* **12** 45-9
[25] Liang A, Zhang X, Yang X and Drury C 2006 Short-term effects of tillage on soil organic carbon storage in the plow layer of black soil in northeast China *Sci. Agr. Sinica* **3** 1287-93
[26] Vita D, Paolo E, Fecondo G, Fonzo N and Pisante M 2007 No-tillage and conventional tillage effects on durum wheat yield, grain quality and soil moisture content in southern Italy *Soil Till. Res.* **92** 69-78
[27] Martinez E, Fuentes J, Silva P, Valle S and Acevedo E 2008 Soil physical properties and wheat root growth as affected by no-tillage and conventional tillage systems in a Mediterranean environment of Chile *Soil Till. Res.* **99** 232-44
[28] Sun G, Chen F, Xiao X, Wu F and Zhang H 2007 Preliminary study on effects of rotational tillage on soil physical properties and rice yield *T. Chin. Soc. Agr. Eng.* **23** 109-13
[29] Han J, Xie J, Zhu J and Wang T 2009 Comprehensive method for treatment of Saline lands *J. Hydraul. Eng.* **40** 372-7
[30] Limon O, Govaerts B, Deckers J and Sayre K 2006 Soil aggregate and microbial biomass in a permanent bed wheat-maize planting system after 12 years *Field Crops Res.* **97** 302-9
[31] The Chinese Academy of Sciences Institute of Nanjing 1983 *Soil Physical and Chemical Analysis* (Shanghai: Shanghai Sci. Press) pp 62-126
[32] Yang P, Luo Y and Shi Y 1993 With the weight of the particle size distribution characterization of soil fractal characteristics *Chin. Sci. Bull.* **38** 1896-9
[33] Qi Y, Wang Y, Liu J, Yu X and Zhou C 2011 Comparative study on composition of soil aggregates with different land use patterns and several kinds of soil aggregate stability index *T. Chin. Soc. Agr. Eng.* **27** 340-7
[34] Nasri B, Fouché O and Torri D 2015 Coupling published pedotransfer functions for the estimation of bulk density and saturated hydraulic conductivity in stony soils *Catena* **131** 99-108
[35] Zhang G, Chan K, Oates A and Huang G 2007 Relationship between soil structure and run off soil loss after 24 years of conservation tillage *Soil Till. Res.* **92** 122-8
[36] Sun B, Wang N and Xie J 2009 Salt and water movement of soil-water body under storage condition *J. Shenyang Agr. U.* **40** 245-7
[37] Han J, Xie J, Wang T, Luo L, Zhu M and Ye X 2009 Experimental observation of Saline alkali of soil in saline land after changing drainage to impoundment in Lubotan of Shaanxi Province *Transactions Chin. Soc. Agr. Eng.* **25** 59-64
[38] Li F, Wang X, Guo Y, Xu X, Yang J, Ke Y and Xiao H 2012 Effect of soil properties and soil enzyme activity in different improvement measures of Saline-alkali soil in Yinchuan plain *Res. Soil Water Conserv.* **19** 13-8
[39] Wu X, Li Y, Li Z, Wang Z and Fan M 2013 Effects of land use types on soil water-stable aggregates in Naban River watershed *Chin. J. Soil Sci.* **44** 1356-63
[40] Lv W, Zhang H, Wang W, Du Z, Wu Y, He P and Xiao L 2010 Characteristics of soil aggregates in different forestlands in simian mountains *Chongqing J. Soil Water Conserv.* **24** 192-7
[41] Shi C, Wang E and Chen X 2010 Organic carbon in soil aggregates and its dynamic changes of larch plantations in black soil region *J. Soil Water Conserv.* **24** 208-12
[42] Rouven A, Koch H and Ludwig B 2014 Effect of long-term tillage treatments on the temporal dynamics of water-stable aggregates and on macro-aggregate turnover at three German sites *Geoderma* **217** 57-64
[43] Choudhury S, Srivastava S, Singh R, Chaudhari S, Sharma D, Singh D and Dipak S 2014 Tillage and residue management effects on soil aggregation, organic carbon dynamics and yield attribute in rice-wheat cropping system under reclaimed sodic soil *Soil Till. Res.* **136** 76-

83
[44] Guo L and Gifford R 2002 Soil carbon stocks and land use change: A meta analysis *Glob. Change Biol* **11** 345-60
[45] Zhao S, Su J, Wu J, Yang Y and Liu N 2006 Changes of soil aggregate organic carbon during process of vegetation restoration in Ziwuling *J. Soil Water Conserv.* **20** 114-7
[46] Six J, Elliott E and Paustian K 2000 Soil macroaggregate turnover and microaggregate formation: A mechanism for C sequestration under no-tillage agriculture *Soil Biol. Biochem.* **32** 2099-103

IOP Conf. Series: Earth and Environmental Science 77 (2017) 012010 doi:10.1088/1755-1315/77/1/012010

Optimization test of the 2BSL-320 vegetable seeders with air-suction drum type

B Tang[1,3], Y S Wang[1] and S Z Ji[2]

[1] Yancheng Vocational Institute of Industry Technology, Yancheng, 224005, China
[2] Jiangsu Cloud Horse Agricultural Machinery Manufacturing Co., Ltd., Yancheng, 224100, China

E-mail: bright9004@126.com

Abstract. The seeding raising technology of the hole tray assembly line is an important part of modern agriculture. The 2BSL-320 vegetable seeders with air-suction drum type are implements that are used to fill nutritional soil and press a hole in a float tray to sow seeds precisely. It can complete the whole process of putting down the tray, bedding the soil, scraping the soil, pressing a hole, sowing the seeds, compacting the soil, watering and putting away the tray by one time. Based on the introduction of the structure and working principle of the implement's critical components, in order to improve the seeding efficiency and the seeding accuracy of the seeders, the response surface tests and the group experiments were carried out in this paper. And the MATLAB tool box was used to conduct fitting and optimization analysis of the test results, also the rationality of the optimization results was validated by experiments, which had provided a theoretical basis for the design of operation parameters in the vegetable seeders and had improved the seeding efficiency and quality.

1. Introduction

With the improvement of agricultural mechanization level, mechanized application of seeding technology has been greatly promoted. As an important part of modern agriculture, the seeding raising technology of the hole tray assembly line has the advantages of high mechanization degree, more crop yield, high seedling survival rate, fewer pests and good crop growth, therefore, it has been applied to the planting and cultivation of some important economic crops such as vegetables and flowers at present.

The structure of seeders used mostly is air-suction type. Due to the characteristics of low seed throwing point, flat soil bed, homogeneous seed distribution, high seeding precision, consistent seed depth and neat emergence etc, which are in line with agronomic requirements, the seeders of air-suction type is becoming more and more popular [1, 2]. Air-suction seeders can be divided into needle type, disc type and drum type, among which the drum type has better seeding speed, precision and efficiency, causing it has a wide range of applications. For the drum seeders, the seeding efficiency is affected by many factors, and how to optimize the design of the relevant factors is the key to improve the efficiency of the drum seeders [3]. This paper introduces the structure and working principle of the 2BSL-320 vegetable seeders with air-suction drum type, and the response surface tests were carried out, after which the test results were fitted and optimized to get the optimal combination of parameters, providing a theoretical basis for the parameters design of the vegetable seeders, improving the seeding efficiency and quality.

2. The structure and working principle

The 2BSL-320 vegetable seeders with air-suction drum type are implements that are used to fill nutritional soil and press a hole in a float tray to sow seeds precisely, which have automatic and manual operation modes. This implement can sow coated and non-coated seeds of tobacco, flowers, vegetables, medicinal herbs in the hole trays, and can not only operate in single line, but also can be used in the standard tray of 50 points, 72 points, 98 points and 128 points which are now widely used at home and abroad. By replacing different drums in the seeder, it can sow many kinds of crops. The overall structure of the implement is shown in figure 1, which is mainly composed of the delivering part, the soil bedding part, the hole pressing part, the sowing part, the soil compacting and fertilization part, the sprinkling part and the electrical control parts, etc. The seeders has different stations, each station can work independently. The complete working process of the seeder is composed of putting down the tray, bedding the soil, scraping the soil, pressing a hole, sowing the seeds, compacting the soil, watering and putting away the tray, in which the rotary power and transmission power are provided by a motor, the power source for the drum absorbing the seeds is from a vacuum pump.

Figure 1. Layout of the seeder.
1-soil bedding part; 2- hole pressing part; 3- sowing part; 4- soil compacting and fertilization part.

2.1. Conveyor

The conveyor is composed of motor, driving sprocket, driving roller shaft, driven roller shaft and conveyor belt etc. Its function is to carry the seed tray according to the designed velocity. In order to monitor the position of the tray on the conveyor belt, the photoelectric reflex zones, as shown in figure 2, are equipped on the delivery path, which use photoelectric principles and can feedback signals when the tray passes the reflection area.

Figure 2. Photo-reflect zone of the conveyor.

Figure 3. Graph of the seeding parts.

2.2. Seeding parts

Seeding part, as shown in figure 3, is the core of the machine, composed of cylinder, vibrator, seed

IOP Conf. Series: Earth and Environmental Science **77** (2017) 012010 doi:10.1088/1755-1315/77/1/012010

box, blowing mouth and so on. The tray with well-pressed holes is delivered to the seeding part by the conveyor belt. There are three main working processes of the seeding part: seeds absorption process, seeds carrying process and seeds dropping process. In the process of absorption, the sucking holes of the cylinder are used to suck the seeds in the seed box due to the vacuum pressure in the cylinder. In the process of carrying, the cylinder rotates carrying the seeds in the sucking hole to the right dropping position. In the process of dropping, a scraping board is used to cut off the vacuum pressure, and then the seeds drop to the tray below due to the effect of the scraping board and self-weight. To facilitate the adsorption of seeds, a vibrator is adopted to make seeds in a state of floating up and down automatically [4].

The effect of seeding was mainly evaluated by the single grain rate, the replay rate and the cavity rate, and the main influencing factors are vacuum pressure, shape of sucking hole, diameter of sucking hole and tangential velocity of cylinder etc. [5, 6].

3. Performance test

3.1. Test condition

The tests are carried out in the 2BSL-320 vegetable seeder with air-suction drum type. The main technical parameters of the machine are shown in table 1.

Table 1. Main technical parameters of the implement.

Model	2BSL-320	Seeding efficiency/reel·hour^{-1}	600
Matched power/W	1080	Seeders type	air-suction
Working breadth/ mm	320	Depth of soil bed/ mm	30~50
Capacity of soil bed/ L	50	Depth of cover soil/ mm	20~35
Capacity of cover soil/ L	50	Compacting depth/ mm	20~25
Irrigation amount / L·reel^{-1}	06-08		

In order to optimize the hole geometry size and the negative pressure parameter of the designed 2BSL-320 vegetable seeder, the hole diameter and the cylinder pressure are chosen as the main optimization parameters, and the single grain rate and the cavity rate are set as the evaluation indexes.

The test subjects are Nanjing cabbage seeds whose diameters are 1.1 ~ 1.5 mm. Because the seed absorption performance of the taper hole is better than that of the straight hole, the shape of the suction hole is tapered. For the two factors of the hole aperture (A) and the cylinder pressure (B), the response surface analysis of the two factors and three levels are carried out. The test factors and levels are shown in table 2.

Table 2. Experiment table of 2 factors and 3 levels.

Levels	Factors	
	Hole diameters (A)/mm	drum pressure(B)/(kPa)
1	1.3	1.5
2	1.6	2
3	2	4

3.2. Test results

The operation process of the seeder tests are: (1) check the condition of the power, water source, gas source and so on (2) sowing the seeds in the seed box (3) power on (4) adjust the compaction height (5) set the operate interface by the manual button (6) put the seeding tray, begin the seeding work by touching the automatic button (7) test over, press the stop key, and then do the clean-up (8) shut off the power, water source and gas source. The group tests are carried out according to the test conditions in table 2, and each test repeats three times, then take down the seed number and cavity number in the tray, and obtain the single grain rate (Y_1) and the cavity rate (Y_2) by calculation. The results of the response surface tests are shown in table 3. It can be seen from table 3 that the values of Y_1 rise firstly and then decrease with the increase of A when B values are the same, and when A values are the same, the change situation of Y_1 is similar. As for the values of Y_2, higher Y_2 values comes up with lower A when B values are the same, while when A values are the same, Y_2 values decrease firstly and then rise with the increase of B.

Table 3. Experiment results of response surface analysis.

Test number	A	B	Y_1/%	Y_2/%	Test number	A	B	Y_1/%	Y_2/%
1	1.3	1.5	82.29	8.14	6	1.6	4	88.76	6.53
2	1.3	2	86.24	6.59	7	2	1.5	84.32	5.08
3	1.3	4	83.69	7.84	8	2	2	87.67	3.37
4	1.6	1.5	86.43	7.45	9	2	4	85.81	4.65
5	1.6	2	91.21	5.23					

4. Test results analysis

The cftool (Curve Fitting Tool) of the MATLAB is powerful and easy to use, and it can implement various types of linear, nonlinear curve fitting. The multivariate regression fitting on the test results in table 2 is carried out by cftool: import the groups of test results in table 2, choose the polynomial fitting, 2 order of fitting. The regression equations of the single grain rate (Y_1) and the cavity rate (Y_2) are obtained respectively as:

$$Y_1= -28.96+110.4A+20.69B-32.75A^2+0.1469AB-3.679B^2$$

$$Y_2= 19.66+1.959A-9.619B-1.94A^2-0.02587AB+1.717B^2$$

The response surface curves of Y_1 and Y_2 with A and B can be seen in figures 4 and 5, respectively.

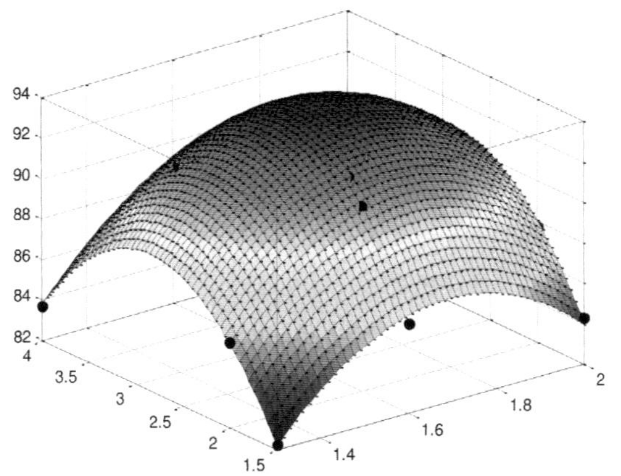

Figure 4. The response surface figure of Y_1 (A, B).

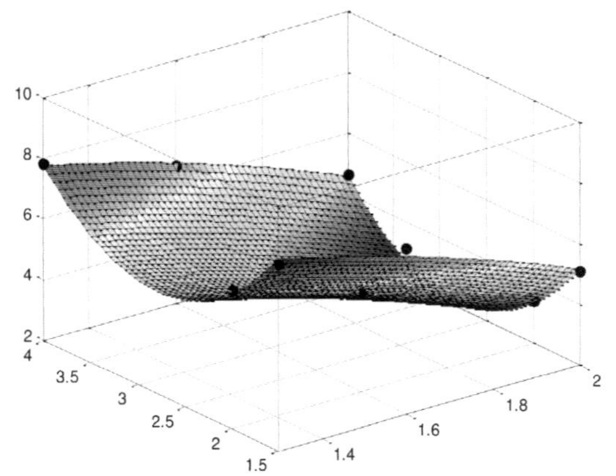

Figure 5. The response surface figure of Y_2 (A, B).

The fitting analysis is taken on the fitting function of the two evaluation index Y1 (A, B) and Y2 (A, B), getting the relevant analysis parameters of fitting are shown in table 4. It can be found from the table 4 that of the two fitting functions, the fitting correlation coefficient (R-square) and adjusted correlation coefficient (Adjust R-square) are close to 1, the variance (SSE) and the root mean square (RMSE) are small, so the correlation of the resulting fitting functions are very good.

Table 4. Parameter list of the fitting analysis.

Data	R-sq	DFE	SSE	RMSE	Adj R-sq
Y_1	0.9916	3	0.4140	0.3715	0.9775
Y_2	0.9951	3	0.0031	0.0323	0.9870

5. Parameters optimization and experimental validation

Optimization problems exist widely in life and many working fields, which are used to discuss the best choice of the decision problem, construct the calculation method in seeking optimal solution. The optimization toolbox in MATLAB can solve many problems of the linear programming, nonlinear programming and multi-objective programming problems and so on [7]. For this reason, optimization toolbox of MATLAB is used to optimize the fitting curve functions obtained, and the ultimate expectation result is maximal single grain rate and minimal cavity rate. The mathematical model is:

- Objective functions: max Y1 (A, B) min Y2 (A, B)
- Constraint conditions: $0.8 \le A \le 2.3$, $1.2 \le B \le 4$

The fgoalattain solver of multi-objective planning is adopted in the optimization process: establish the objective function of m file, make calls of the m file in the fitting toolbox and input the constraints. The resulting optimization results are:

$$A=1.983,\ B=2.816,\ Y_1=91.0901,\ Y_2=2.3001$$

The actual tests are taken in the 2BSL-320 vegetable seeder with the result of $Y_1 = 89.42$, $Y_2 = 2.93$ through calculation processing to validate the above optimization results, which is in accordance with the theory results. It indicates that the optimization results can provide certain theoretical basis for the design of the operation parameters in the vegetable seeder.

6. Conclusions

The main job of this paper includes:

- Introduce the structure, and working principle about the 2BSL-320 vegetable seeder with air-suction drum type in details, and carry out the seeder tests based on a response surface analysis of the two factors and three levels with the two factors of the aperture (A) and the cylinder pressure (B), the single grain rate and the cavity rate as the evaluation indexes.
- Implement a multivariate regression fitting on the test results by cftool of the MATLAB, and obtain the regression equations of the single grain rate and the cavity rate, on which the multi-objective optimization is based through the optimization toolbox. The resulting optimization results are: $A=1.983$, $B=2.816$, $Y_1=91.0901$, $Y_2=2.3001$.
- Accomplish the actual tests in the 2BSL-320 vegetable seeder with the parameters of $A=1.983$, $B=2.816$ to validate the optimization results, in which the final results is $Y_1 = 89.42$, $Y_2 = 2.93$ that is in accordance with the theory results.

This paper can provide certain theoretical basis for the design of the operation parameters about the 2BSL-320 vegetable seeder.

Acknowledgements

This work was sponsored by the special funds on the northern Jiangsu science and technology of Jiangsu Province: the development and industrialization of the vegetable seeder (BN2014059).

IOP Conf. Series: Earth and Environmental Science **77** (2017) 012010 doi:10.1088/1755-1315/77/1/012010

References

[1] Gaikwad B B and Sirohi N P S 2008 Design of a low-cost pneumatic seeder for nursery plug trays *Biosyst. Eng.* **99** 322–9

[2] Zhao X S, Yu H L, Zhang J G and Zhang B 2016 Pneumatic precise radish seeder *Adv. J. Food Sci. Technol.* **10** 180–4

[3] Movahedi E, Rezvani M and Hemmat A 2016 Design, development and evaluation of a pneumatic seeder for automatic planting of seeds in cellular trays *Trans. Chin. Soc. Agric. Mach.* **4** 65–72

[4] Abdolahzare Z, Asoodar M A, Kazemi N, Rahnama M and Mehdizadeh S A 2016 Optimization of the most important operational parameters of a pneumatic seeder using real-time monitoring for cucumber and watermelon seeds *Trans. Chin. Soc. Agric. Mach.* **6** 35–48

[5] Karayel D, Barut Z B and özmerzi A 2004 Mathematical modelling of vacuum pressure on a precision seeder *Biosyst. Eng.* **87** 437–44

[6] Yazgi A and Degirmencioglu A 2007 Optimisation of the seed spacing uniformity performance of a vacuum-type precision seeder using response surface methodology *Biosyst. Eng.* **97** 347–56

[7] Chen J 2011 *MATLAB Book* vol 3 (Beijing: Publishing House of Electronics Industry) pp 158–214

Fertilizer nitrogen, soil chemical properties, and their determinacy on rice yield: Evidence from 92 paddy fields of a large-scale farm in the Kanto Region of Japan

D Li[1], T Nanseki[1, 3], Y Chomei[1] and S Yokota[2]

[1] Faculty of Agriculture, Kyushu University, 812-8581, Hakozaki 6-10-1, Fukuoka, Japan
[2] Yokota Nojo Co. Ltd., 301-0803, Nuruto 2047, Ryugasaki, Ibaraki, Japan

E-mail: nanseki@agr.kyushu-u.ac.jp

Abstract. Rice, a staple crop in Japan, is at risk of decreasing production and its yield highly depends on soil fertility. This study aimed to investigate determinants of rice yield, from the perspectives of fertilizer nitrogen and soil chemical properties. The data were sampled in 2014 and 2015 from 92 peat soil paddy fields on a large-scale farm located in the Kanto Region of Japan. The rice variety used was the most widely planted Koshihikari in Japan. Regression analysis indicated that fertilizer nitrogen significantly affected the yield, with a significant sustained effect to the subsequent year. Twelve soil chemical properties, including pH, cation exchange capacity, content of pyridine base elements, phosphoric acid, and silicic acid, were estimated. In addition to silicic acid, magnesia, in forms of its exchangeable content, saturation, and ratios to potassium and lime, positively affected the yield, while phosphoric acid negatively affected the yield. We assessed the soil chemical properties by soil quality index and principal component analysis. Positive effects were identified for both approaches, with the former performing better in explaining the rice yield. For soil quality index, the individual standardized soil properties and margins for improvement were indicated for each paddy field. Finally, multivariate regression on the principal components identified the most significant properties.

1. Introduction

Soil is critical to crop growth because it provides a growth environment and indispensable nutrients, and any degradation of soil quality may result in decreased productivity, quality, and thus profitability of the crop [1-4]. Soil properties can be determined primarily from physical, chemical, and biological aspects [5, 6]. The chemical properties typically relate more directly to the sustainability of the agro-ecosystem, in addition to variability in crop yield [2, 6, 7]. These properties can also be more easily improved than others, through proper fertilization and other farm management practices [8]. Thus, soil fertility may also refer to its chemical properties, with different aspects taken into consideration. Many researchers have analyzed the relationship between soil chemical prosperities and rice yield. Juhos *et al* [1] constructed a soil quality index using three principal components derived from more than ten basic indicators to evaluate the chemical and physical properties of soil and the impact on yields of maize, winter wheat, and sunflower, sampled from a 225 ha farmland in east Hungary. Liu *et al* [3] analyzed the rice yield and the effects of eight soil chemical properties sampled from 13 provincial regions of south China, individually and synthetically using a soil quality index, based on a principal component analysis model adapted from that published by Qi *et al* [8].

In Japan, rice (Oryza sativa L.) is the most important staple crop, and by 2015, it accounted for the largest proportion of 17% of gross agriculture output [9]. Japan is striving to improve rice productivity and global competitiveness. By 2016, the total planted area of rice was estimated to be 1.57 million ha, having decreased by approximately one third over the past three decades. At the same time, the total rice production decreased by nearly 30% [10]. Rice yield depends more on soil fertility than other crops, and loss of the crop may result from unbalanced soil fertility arising from improper fertilization and the nutrient deposited in the dammed rivers [5]. Thus, accurate measurement of soil fertility and its effect on yield is essential to promote rice production in Japan. Some scholars have investigated the chemical properties of paddy field soil in Japan. To estimate total CH_4 emission from rice paddies, Katayanagi *et al* [11] analyzed the soil chemical properties of 986 plots sampled across Japan, using the individual indicators of pH (H_2O) and total carbon. Matsumoto *et al* [12] estimated the effects of iron materials applied in an experimental field at Shimane University, Matsue city of Shimane Prefecture, and determined soil chemical properties in terms of the content of available arsenic and phosphorus and acid ammonium oxalate extractable iron and aluminum. To measure the effect of fermented bark as a soil amendment material in the experimental site of Gunma Prefecture, Japan, Mori *et al* [13] determined the soil chemical properties using pH, cation exchange capacity (CEC), oxidation-reduction potential, and the content of heavy metals, including cadmium, copper, and zinc. Thus, it is necessary to quantify the soil chemical properties using synthesized indices, and on-farm data from individual paddy fields can provide more practical information.

Here, we discuss the status of fertilizer nitrogen and soil chemical properties in the sampled paddy fields and demonstrate the effect of fertilizer nitrogen and soil chemical properties on rice yield, through the determination of soil quality index and standardized soil properties. We also examine the most significant properties using principal component analysis and multivariate regression, and summarize empirical findings and as well as countermeasures to improve rice yield.

2. Materials and methods

2.1. Sampling and data

The data were sampled in 2014 and 2015 from 92 peat soil paddy fields on a large-scale farm located in the Kanto Region of Japan. The rice variety was Koshihikari, the taste and appearance of which make it desirable to consumers both in Japan and worldwide. By 2015, it accounted for the largest share (36%) [14] of the area planted with domestic rice, partly due to its strong cold resistance and stable yield [15]. Rice yield was measured as the grain weight of paddy with 15% moisture content. The weight and moisture content of the raw paddy were monitored by combine harvesters equipped with a matchbox-sized sensor set at the input slot of the grain tank, and the global navigation satellite system (GNSS) to convey data to the cloud server shared by companies, institutes, and farms.

Fertilizer nitrogen was calculated based on the amounts of chicken manure, chemical fertilizer, ammonium sulfate, and urea fertilizers applied and their corresponding nitrogen content. In accordance with local guidelines [16], we presented the soil chemical properties in terms of 12 indicators. The pH specifies the acidity (<7) or basicity (>7). CEC is a measure of soil fertility, i.e., the capacity of soil to hold positively charged ions of NH_4^+, Ca_2^+, Mg_2^+, and K^+, protecting the groundwater from cation contamination [5, 13]. Phosphoric acid is essential to ensure grain quality, while excessive content can lead to premature or low yield [17]. Silicic acid is indispensable for rice growth by preventing the softening of stems and leaves, and root decay. Contents and ratios of the pyridine base elements include lime that is indispensable for root growth, magnesia necessary for photosynthesis, and potassium essential for anthesis and seed-setting [5].

2.2. Analysis framework

Firstly, we analyzed the effect of fertilizer nitrogen through regression analysis of rice yield over two years. We analyzed the effects of the soil chemical properties by constructing a soil quality index (SQI). Using the standardized SQI for each paddy field, we identified relative margins to improve the

IOP Conf. Series: Earth and Environmental Science **77** (2017) 012011 doi:10.1088/1755-1315/77/1/012011

soil properties. We also conducted a principal component analysis on the 12 soil chemical properties, extracted five principal components of the total variance, and analyzed their effects on rice yield. Finally, we identified the most significant properties using multivariate regression on the principal components. The regression and principal component analyses were performed using SPSS 23.0 for Windows (IBM Corp.).

3. Results and discussion

3.1. Effect of fertilizer nitrogen

The amount of fertilizer nitrogen affected yield significantly and positively, according to the results observed over the two years (figure 1(a)). In 2015, the linear relationship between fertilizer nitrogen and rice yield was significant, while it was non-significant in 2014. This indicated that fertilization was improved, thereby increasing rice yield across the sampled paddy fields. Meanwhile, as illustrated by the scatter plot, the amount of fertilizer nitrogen was greater in 2015 (69 kg ha^{-1}) than in 2014 (50 kg ha^{-1}). As a result, a quadratic relationship was estimated to be significant in 2015, indicating diminishing returns when the fertilizer nitrogen exceeded approximately 105 kg ha^{-1}.

We hypothesized that it might take some time for the fertilizer nitrogen to enrich the soil and thus improve plant growth. Hence, we identified the significant and positive effects of the amount of fertilizer nitrogen in 2014 on the rice yield of 2015. On average, a 1 kg increase in fertilizer nitrogen in 2014 resulted in 12.8 kg of paddy yield per hectare in 2015. This was higher than the effect of the fertilizer nitrogen in 2015, which was merely 8.3 kg (figure 1(b)). The results confirmed the sustained nature of the effect of fertilizer nitrogen on rice yield; it is therefore necessary to investigate the amount of residual nitrogen in the soil before fertilization [17]. Over the two years, ammonium sulfate fertilizer accounted of the largest ratio of fertilizer nitrogen by 40.8%, following by manure fertilizer (40.3%) and chemical compound fertilizer (18.9%).

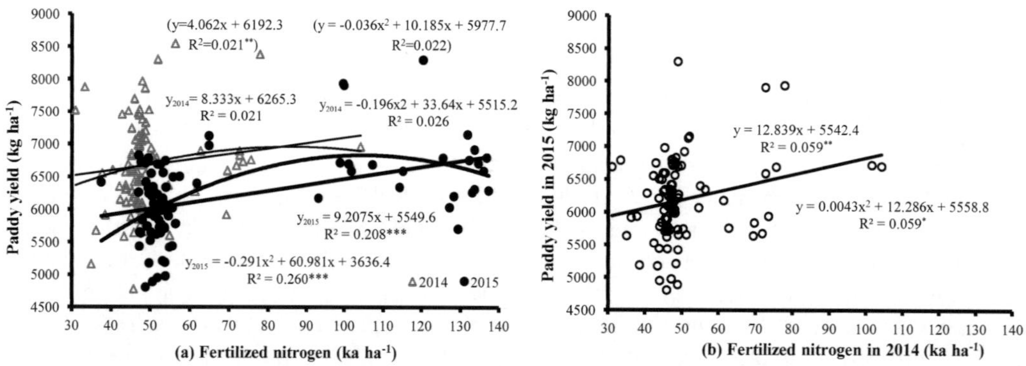

Figure 1. Relationship between (a) fertilizer nitrogen and rice yield in 2014-2015, and (b) fertilizer nitrogen in 2014 and rice yield in 2015.
Note: ***, ** and * indicate significant at 1%, 5% and 10%, respectively.

3.2. Effects of the individual soil properties

In Japan, the soil tends to be acidic owing to a substantial amount of precipitation, which washes away the alkaline calcium and magnesium components [17]. The CEC of the sampled fields was slightly lower than the local standard, indicating that the soil fertility needed to be improved. The negative correlation with rice yield may indicate that, in Japan, phosphoric acid can be more easily fixed by rich volcanic ash soil and supplied through overused organic fertilizers [5]. Higher content of silicic acid increased yield, and a positive correlation was observed between them (table 1). Of the pyridine base elements, significant and positive effects were observed for magnesia, from the perspectives of its exchangeable content, saturation, and its ratios to potassium and lime (table 1).

Table 1. Rice yield, fertilizer nitrogen and soil chemical properties of 92 peat soil paddy fields in 2014-2015. Each property is evaluated based on the average and ideal values.

Variable	N	Mean	CV [a] (%)	Optimum	Score [b]	Corr. [c]
Paddy yield of 15% moisture (kg ha^{-1})	184	6434.38	10.86	—	—	1.000
Fertilizer nitrogen (kg ha^{-1})	184	59.61	42.27	—	—	0.150**
pH	184	6.16	2.93	6	1.03	-0.072
CEC (meq 100 g^{-1})	184	18.61	30.25	27	0.69	0.015
Effective phosphoric acid (mg 100 g^{-1})	184	9.61	54.99	10~30	0.96	-0.289***
Effective silicic acid (mg 100 g^{-1})	184	20.75	55.93	30~40	0.69	0.414***
Exchangeable potassium (mg 100 g^{-1})	184	19.14	29.04	25~30	0.77	0.055
Exchangeable lime (mg 100 g^{-1})	184	303.71	28.71	300~350	1.00	0.033
Exchangeable magnesia (mg 100 g^{-1})	184	58.43	31.80	35~40	1.46	0.291***
Potassium saturation (%)	184	2.31	31.58	2.0~2.5	1.00	0.013
Lime saturation (%)	184	58.82	12.23	40~45	1.31	0.015
Magnesia saturation (%)	184	16.04	26.14	6~7	2.29	0.317***
Lime/magnesia	184	3.85	22.30	5.4~7.1	0.71	-0.347***
Magnesia/potassium	184	7.52	37.78	2.7~3.8	1.98	0.209***

[a] Coefficient of variance.
[b] Equals to 1 when the mean falls into the optimal range, or dividing the mean by the corresponding nearer bound, lower or upper, of the optimal range.
[c] Correlation coefficient for yield determined using paddy of 15% moisture content in 2014-2015, while *** and ** * indicate significant correlation at 1% and 5%, respectively.
Data source: survey by the authors conducted in 2014-2015, and the optimum was MAFF [9].

3.3. *Soil quality index*

Based on to the Hungarian soil quality index adopted by Juhos *et al* [1], and using the 12 chemical properties discussed above, we determined the SQI for each paddy field i as:

$$SQI_i=\sum_{j=1}^{12} SSP_i \times w_j, \quad SSP_i=\sum_{j=1}^{12}\frac{SP_{ij}-\min(SP_j)}{\max(SP_j)-\min(SP_j)}, \quad w_j=R_j/\sum_{j=1}^{12}|R_j| \quad (i=1, 2, \ldots, n), \tag{1}$$

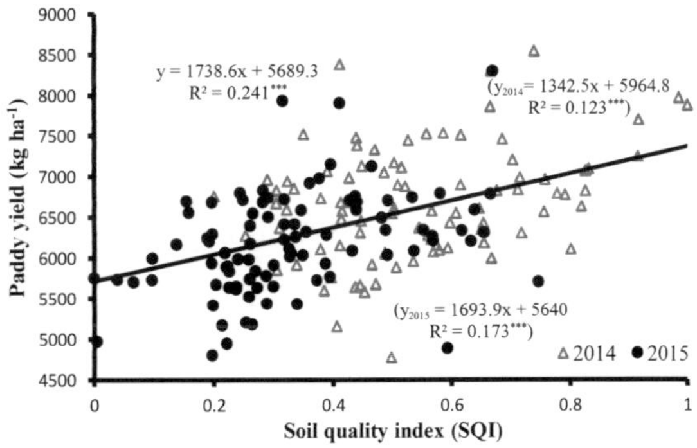

Figure 2. Regression analysis between the rice yield and the SQI in 2014-2015.
Note: *** indicates significant correlation at 1%.

where SSP is the standardized soil property; w_j is the weight of soil property j (SP_j); $|R_j|$ is the absolute value of correlation coefficient of SP_j with the paddy yield, as indicated in table 1; and n is the number of paddy fields. To calculate the relative position between the corresponding minimum and maximum values of the paddy fields (equation (1)), we standardized both the soil property (SSP) and the SQI from 0 to 1. A significant correlation was observed between SQI and rice yield (figure 2).

The determination coefficient (R^2) indicated that 24% of the yield was explained by the soil quality defined in equation (1). This result thus supported the hypothesis that good soil chemical properties are essential to increase rice yield [3]. In addition, comparison of the SQIs demonstrated the differences in soil quality among the sampled paddy fields. In this study, the two-year average SQI was 0.536, which was higher than the SQI of 0.323 observed in 2015. This might have contributed to the reduced average rice yield per hectare, from 6683 kg to 6186 kg, over the two years. However, with respect to its correlation with rice yield, the determination coefficient R^2 in 2015 was higher than that in 2014, indicating that soil quality had a higher contribution to rice yield in 2015 (figure 2).

Using indicator-specific SSPs for each paddy field, we determined relative soil quality based on different aspects, for use as a standard for soil quality improvement through optimal fertilization and other management strategies. In the paddy fields with SQI values of 1 and 0, the highest SSPs were observed for the exchangeable magnesia (figure 3(a)) and the ratio of lime to magnesia (Figure 3(b)), while the SSPs were more balanced in paddy fields with average SQI values (Figure 3(c)). Similar radar charts were available for all the other paddy fields.

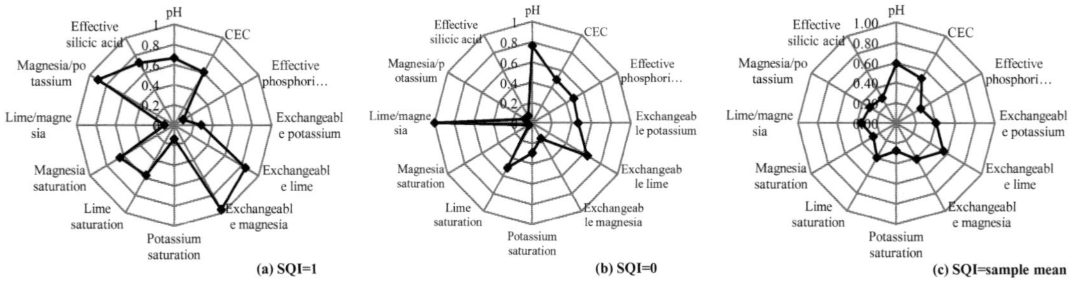

Figure 3. Relative soil quality of paddy fields with different standardized soil properties (SSPs).
Note: *** indicates significance at 1%.

3.4. *Principal component analysis (PCA)*

PCA has been used in several studies to identify and separate soil quality indicators [1, 2, 7]. In this study, we conducted the PCA on the 12 soil chemical properties with varimax rotation. Using a cutoff of over 1 for the eigenvalues indicating components variances, we extracted five principal components, which explained 91.34% of the total variance. The KMO measurement (0.61) of sample adequacy and Bartlett's test of sphericity (sig. at 0.01) indicated that the PCA was appropriate [18].

Table 2. Rotated component matrices of five principal components of the soil quality indicators.

Soil quality indicator	PC1	PC2	PC3	PC4	PC5
pH (SP1)	0.103	0.192	0.177	**0.784**	-0.112
CEC (meq 100 g^{-1})	**0.952**	-0.152	-0.036	-0.206	-0.104
Effective phosphoric acid (mg 100 g^{-1})	0.136	-0.096	0.05	0.053	**-0.901**
Effective silicic acid (mg 100 g^{-1})	0.235	0.282	0.161	0.309	**0.713**
Exchangeable potassium (mg 100 g^{-1})	**0.567**	-0.062	**0.805**	0.032	0.057
Exchangeable lime (mg 100 g^{-1})	**0.959**	-0.175	-0.064	0.159	0.011
Exchangeable magnesia (mg 100 g^{-1})	**0.799**	**0.527**	-0.127	0.143	0.148
Potassium saturation (%)	-0.506	0.097	**0.795**	0.186	0.105
Lime saturation (%)	-0.08	-0.014	-0.081	**0.905**	0.24
Magnesia saturation (%)	-0.134	**0.819**	-0.119	0.445	0.229
Lime/magnesia	0.125	**-0.971**	0.037	0.046	-0.133
Magnesia/potassium	0.282	**0.543**	**-0.755**	0.085	0.102
Explained variance after rotation (%)	27.069	19.821	16.275	15.526	12.653
Cumulated %	27.069	46.890	63.166	78.692	91.344

KMO measurement of sample adequacy: 0.610; Bartlett's test of sphericity: Chi-Square (66) = 019.791***.
Rotation method: varimax with Kaiser normalization converged in six iterations, and the high factor loadings are bolded.
Software: IBM SPSS 23.0 for windows.

Principal component 1 (PC1) was identified as CEC and content of exchangeable pyridine base elements, due to their high loadings. PC1 accounted for 27.07% of the total variance. Similarly, the other PCs were labeled as magnesia (PC2), potassium (PC3), pH and lime (PC4), and phosphoric and silicic acids (PC5), respectively, considering their high loadings of the related properties. Accordingly, the variance explained by these PCs decreased from 19.82% to 12.65% (table 2). Weighting the five principal components using the corresponding percentage of variance explained, we determined the principal component and regressed it with the paddy yield. As shown in figure 4(a), the determination coefficient (R^2 = 0.092) was significant, but less than that of the regression with SQI as the independent variable (figure 2). Similar results were obtained with the models of the other year. Thus, the SQI performed better as an indicator for increase in rice yield.

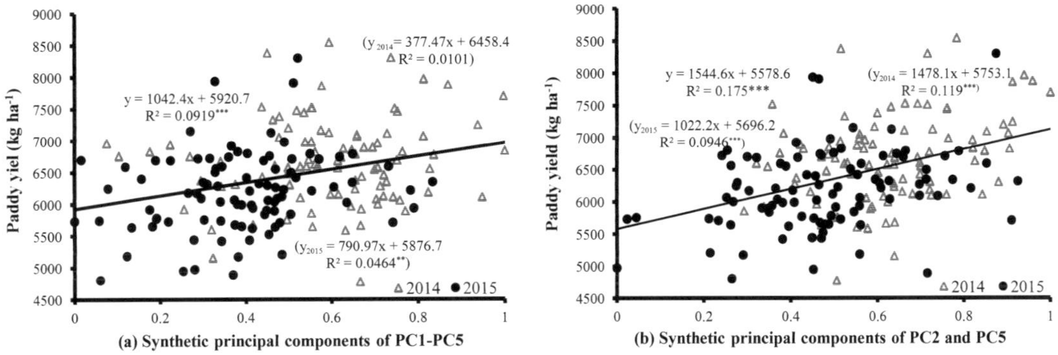

Figure 4. Synthetic scores of five principal components (SPC) and rice yield in 2014-2015.
Note: *** and ** indicate significance at 1% and 5%, respectively.

To explore the possible reasons, we conducted principal component regression analysis [1, 4], using the stepwise method to select the variable. The result indicated that only PC2 and PC5 were significant (table 3). Accordingly, as illustrated in figure 4(b), the determination coefficients increased when regression analysis was performed on the determined principal components of PC2 and PC5. Thus, integrating with the respective high loadings, bold values in table 2, the significant properties included magnesia and phosphoric and silicic acids contents.

Table 3. Results of multivariate regression on the five principal components.

Independent variable	Unstandardized coefficient		Standardized coefficient	t	Sig	Collinearity statistics	
	B	Std. E	Beta			Tolerance	VIF
PC2	227.622	45.383	0.326	5.016	0.000	1.000	1.000
PC5	252.772	45.383	0.362	5.570	0.000	1.000	1.000
(Constant)	6434.376	45.259	0.362	142.167	0.000	1.000	1.000
N = 184, R = 0.487, R^2 = 0.237, Adj. R^2 = 0.228, F (2, 181) = 28.089***							

Dependent variable: paddy yield at 15% moisture content (kg ha^{-1}).
Independent variables selecting method: stepwise out of PC1 through PC5; *** indicates significance at 1%.
Software: IBM SPSS 23.0 for windows.

4. Conclusions

The amount of fertilizer nitrogen affected the yield significantly and positively over the two years, while a turning point was observed when the fertilizer nitrogen was approximately 105 kg ha^{-1}. The sustained nature of the effect of fertilizer nitrogen was confirmed. Thus, the effects of residual nitrogen on rice yields in the subsequent years need to be investigated to ensure proper fertilization.

Based on the 12 chemical properties, the constructed SQIs were found to be significantly related to the rice yield. The higher determination coefficient R^2 in 2015 indicated that soil quality was more important in explaining rice yield than that in 2014. Using the SSP of each paddy field, we identified the relative soil quality in terms of different properties, providing a reference for soil quality improvement through improved fertilization and other management practices. Comparison of the paddy fields with the highest and lowest SSPs demonstrated that magnesia was an essential soil indicator. Similarly, PCA and stepwise multivariate regression indicated that the most significant soil properties included magnesia and phosphoric and silicic acid content. The SQIs and SSPs provided important indices to monitor both the overall and individual soil properties of paddy fields. When considering measures to improve rice yield, SQIs and SSPs should be included in the panel database of rice yield and soil properties. Thus, the analysis framework can be adapted for multiple farms of different regions and soil types, incorporating several more soil properties.

Acknowledgments

This study was supported by the Cabinet Office, Government of Japan, Cross-ministerial Strategic Innovation Promotion Program (SIP), "Technologies for creating next-generation agriculture, forestry, and fisheries" (funding agency: Bio-oriented Technology Research Advancement Institution, NARO).

The authors thank Dr. Yuji Matsue, Kyushu University, and the anonymous reviewers, for their kind comments and support in preparing this paper. Any remaining errors are due to the authors.

References

[1] Juhos K, Szabób S and Ladányic M 2016 Explore the influence of soil quality on crop yield using statistically-derived pedological indicators *Ecolo. Indic.* **63** 366-73

[2] Li Y, Chen X, Shamsi I H, Fang P and Lin X 2012 Effects of irrigation patterns and nitrogen fertilization on rice yield and microbial community structure in paddy soil *Pedosphere* **22(5)** 661-72

[3] Liu Z, Zhou W, Shen J, He P, Lei Q and Liang G 2014 A simple assessment on spatial variability of rice yield and selected soil chemical properties of paddy fields in South China *Geoderma*. **235-236** 39-47

[4] Obade V P and Lal R 2016 A standardized soil quality index for diverse field conditions *Sci. Total Environ.* **541** 424-34

[5] DFPJA (Department of fertilizer and pesticide, National federation of agricultural cooperative associations of Japan, JA Zen-Noh) 2011 *Guide to Soil Diagnosis and Fertilizer Calculation* (Tokyo, Japan: Rural Culture Association Press) pp 12-3, 16, 72-7 (in Japanese)

[6] Bouma J 2002 Land quality indicators of sustainable land management across scales *Agric. Ecosyst. Environ.* **88** 129-36

[7] Qi Y, Darilek J L, Huang B, Zhao Y, Sun W and Gu Z 2009 Evaluating soil quality indices in an agricultural region of Jiangsu Province, China *Geoderma* **149** 325-34

[8] Gray L C and Morant P 2003 Reconciling indigenous knowledge with scientific assessment of soil fertility changes in southeast Burkina Faso *Geoderma* **111** 425-37

[9] MAFF 2016 a Statistics on incomes from agricultural production, http://www.maff.go.jp/j/tokei /kouhyou/nougyou_sansyutu/attach/pdf/index-2.pdf (in Japanese, accessed 2017.2.20)

[10] MAFF 2017 Estimated planted area and production of paddy rice, http://www.maff.go.jp/j/ press/tokei/seiryu/pdf/141030-01.pdf (in Japanese, accessed 2017.1.18)

[11] Katayanagi N, Fumoto T, Hayano M, Takata Y, Kuwagata T, Shirato Y, Sawano S, Kajiura M, Sudo S, Ishigooka Y and Yagi K 2016 Development of a method for estimating total CH4 emission from rice paddies in Japan using the DNDC-Rice model *Sci. Total Environ.* **547** 429-40

[12] Matsumoto S, Kasuga J, Makino T and Arao T 2016 Evaluation of the effects of application of iron materials on the accumulation and speciation of arsenic in rice grain grown on uncontaminated soil with relatively high levels of arsenic *Environ. Exp. Bot.* **125** 42-51

[13] Mori M, Kotaki K, Gunji F, Kubo N, Kobayashi S, Ito T and Itabashi H 2016 Suppression of cadmium uptake in rice using fermented bark as a soil amendment *Chemosphere* **148** 487-94

[14] Komenet 2016 Planting areas of the major Japonica varieties in Japan, http://www. komenet.jp/jukyuudb/826.html (in Japanese, accessed 2017.2.17)

[15] Goto Y, Nitta Y and Nakamura S 2000 *Crops I: Paddy Cultivation* (Tokyo Japan: Agricultural Development and Extension Association press) 210-1 (in Japanese)
[16] MAFF 2016 b Soil improvement standards (Ibaraki), http://www.maff.go.jp/j/seisan/kankyo/hozen_type/h_sehi_kizyun/attach/pdf/ibaraki01-3.pdf (in Japanese, accessed 2016.12.22)
[17] Fujiwara S, Anzai T and Kato T 1996 *Approaches and Application of Soil Diagnosis* (Tokyo, Japan: Rural Culture Association Press) pp 23, 26-7, 88 (in Japanese)
[18] Hutcheson G D and Sofroniou N 1999 *The Multivariate Social Scientist: An Introduction to Generalized Linear Models* (Thousand Oaks, California: Sage Publications) pp 224-5

Hydrodynamic characteristics of the oval cambered double slotted otter board in bottom trawl fisheries

J Liu[1,2,3], H L Huang[1,2,3,4], L Z Li[1,2,3], T C Qu[1,2,3], Y Wu[1,2,3], S Chen[1,2,3], J L Yang[1,2,3] and X Rao[1,2,3]

[1] Engineering Technology Research Center of Marine Fishing, Chinese Academy of Fishery Sciences, Shanghai 200090, China
[2] Key Laboratory of East China Sea and Oceanic Fishery Resources Exploitation, Ministry of Agriculture, Shanghai 200090, China
[3] East China Sea Fisheries Research Institute, Chinese Academy of Fishery Sciences, Shanghai 200090, China

E-mail: huanghl@ecsf.ac.cn

Abstract. The otter board is one of the main components of single boat trawl fisheries. An oval cambered double slotted otter board was developed for improving the expansion performance of trawl net in bottom trawl fisheries. A flume model experiment was conducted to measure the lift coefficient (C_L), drag coefficient (C_D), and lift to drag ratio (K) in different angles of attack (α). The experimental results are as follows: (1) The C_L and K value show a trend of increasing at the beginning and then decreasing with the increase of angle of attack, the C_D value reflects an upward trend as the angle of attack increases; (2) The D3 otter board (front flow deflector angle at 29°) showed a better hydrodynamic performance. When α=30°, the max lift coefficient (C_{Lmax}) was 1.464, in this case C_D =0.554 and K=2.643. When α=15°, the max lift to drag ratio (K_{max}) was 4.165, C_L =0.633, and C_D =0.152. This suggests that the best working scope for the angle of attack is between 15°~30°, in which case, C_L>0.633 and K>2.643. The mean value of the lift coefficient was 1.071 and the mean of the lift to drag ratio was 3.482. Comparative analysis of the hydrodynamic performance of different types of otter boards showed that the D3 otter board both had good expansion performance and efficiency, which can provide a reference basis for further optimization of the bottom trawl otter board.

1. Introduction

Globally, single boat trawling is responsible for nearly 40% of marine fishing [1]. The main equipment used in single trawl fishing vessels is the otter board, and its hydrodynamic performance is an important factor in the efficiency of catching and fishing [1-5]. The otter board both accelerates the settlement of the trawl net and increases the horizontal expansion of the trawl net [2, 3]. The otter board used in distant fishing boat has various structural styles, including vertical cambered, vertical cambered slotted, and vertical cambered V type [2, 3]. The V type otter board and oval cambered otter board are mostly used in near shore single boat trawl fisheries [2, 6]. The hydrodynamic force on the otter board decomposes into the expansion force and water resistance. The expansion force is perpendicular to the flow velocity, which has the effect of increasing the trawl net sweeping area. Water resistance is parallel to the flow of water. The ratio of the expansion force and water resistance is an important parameter for measuring the expansion efficiency of the otter board. The hydrodynamic performance of otter board has been studied through two methods: model testing or numerical simulation [6-10]. The model test in a circulating water tank closely approximates the actual working conditions of trawl fishing [5]. Japanese scholars have analyzed the hydrodynamic performance of otter board through the flume model experiment [11-14]. Chinese scholars have focused on analysis of the hydrodynamic performance of the otter board in wind tunnels [1, 15-17], and in recent years have used the flume model experiment [18-21].

Developed countries have sought to improve the hydrodynamic performance of the otter board. Currently, the maximum efficiency of an oval cambered single slotted otter board is 1.25, with a lift coefficient of 0.93, and a drag coefficient of 0.74 [3]. To improve the expansion performance of trawl nets in bottom trawl fisheries, an oval cambered double slotted otter board model experiment is carried out and the effect of the change of the basic parameters (flap angle) on the hydrodynamic performance is studied. The analysis of the hydrodynamic performance of different types of otter boards can provide a reference point for further optimizing the bottom trawl otter board.

2. Materials and methods

2.1. Experimental instruments

The model test was conducted in a circulating water tank at the East China Sea Fisheries Research Institute. The scale of the flume experiment was 180 cm × 50 cm × 50 cm, with a maximum flow rate of 2.5 $m \cdot s^{-1}$. The experimental device is shown in figure 1. The experimental model was installed in the middle portion of the flume experiment, which connected with the three-component force sensor through the connecting rod. The three-component force sensor fixed on a rotary table used for the machine tool. The model's angle of attack could be changed by adjusting the rotary table. The measuring instrument was a three-component force sensor LSM-B-500NSA1-P made by Kyowa Co., Ltd., Japan. The measuring range was 500 N. Data was derived from the computer.

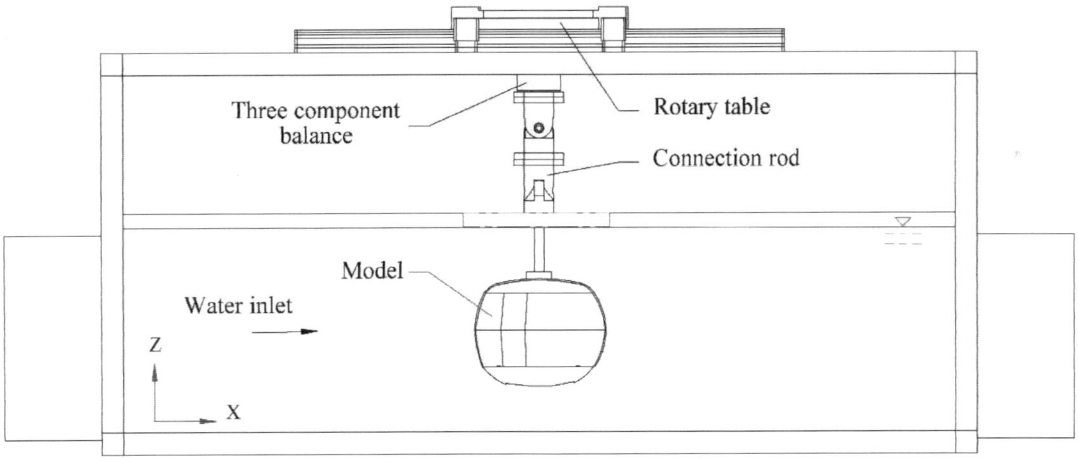

Figure 1. Schematic diagram of circulating water tank with test model.

2.2. Experimental models

Three series models were produced using the different front flow deflector angles of oval cambered double slotted otter boards, as shown in table 1. The model scale was 1:20 and the wall thickness was 3 mm (figure 2). The top of the experimental model was connected with a support through a M4 threaded hole. The screen model is made of photosensitive resin composite material. The process parameters were as follows, model precision: ±0.1 mm, detail resolution: 0.4 mm, minimum wall thickness: 1 mm, print layer thickness: 0.1 mm. The model used a method of stereo light curing (3D print technology) derived from a computerized model, which has the advantage of high strength and resistance to deformation. Compared with the traditional method (manufacturing from stainless steel), the 3D print model has the advantages of high precision, a short manufacturing cycle, low cost, and suitability for highly complex modeling. It also has the advantages of higher accuracy and better surface finish than the fused deposition modeling method.

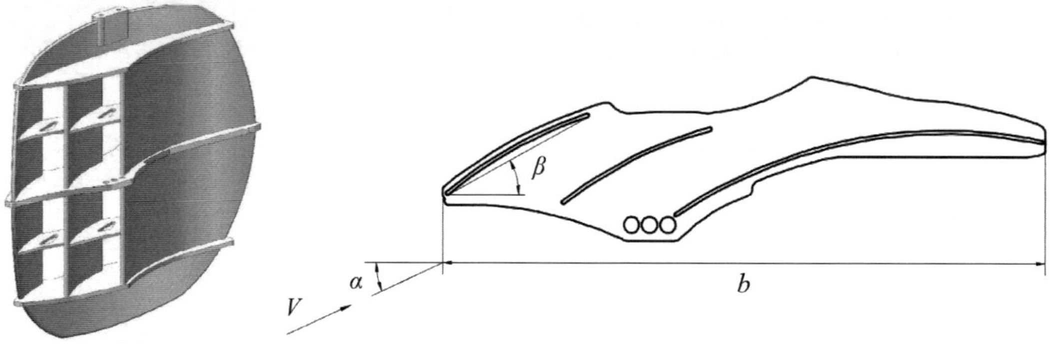

(a) (b)

Figure 2. Experimental otter board (a) and diagram of its cross section (b).

Table 1. Specification of model otter board.

No.	Front flow deflector angle β	Parameters	Chord length b (cm)	Aspect ratio λ	Flow area (m^2)
D1	23°	Middle flow deflector angle 26°; Back flow deflector angle 11°	26	0.8	0.048
D2	26°				
D3	29°				

2.3. Experimental conditions

The factors and levels are shown in table 2. The interval is 2.5° between 15°~40°, and 5° between 0°~15° and 40°~60°. The experimental flow velocity range is 0.6~1.0 $m\cdot s^{-1}$ with an interval of 0.1 $m\cdot s^{-1}$.

Table 2. Experimental conditions.

Factor	Level
Angle of attack α	0°, 5°, 10°, 15°, 17.5°, 20°, 22.5°, 25°, 27.5°, 30°, 32.5°, 35°, 37.5°, 40°, 50°
Flow velocity ($m\cdot s^{-1}$)	0.6, 0.7, 0.8, 0.9, 1.0

2.4. Data processing

The aspect ratio (λ) is defined as the ratio of the span length l and chord length b. The results of the resistance F_x and F_y were recorded using a three-component balance. The drag coefficient C_D, lift coefficient C_L and lift to drag ratio K were calculated after pole interference correction. The calculation equations follow:

$$C_D = \frac{F_x}{0.5\rho V^{2} S} \tag{1}$$

$$C_L = \frac{F_y}{0.5\rho V^{2} S} \tag{2}$$

$$K = \frac{C_L}{C_D} \tag{3}$$

$$Re = \frac{Vb}{\upsilon} \tag{4}$$

In the equations, ρ stands for fluid density; V stands for flow velocity ($m\cdot s^{-1}$); S stands for area (m^2); Re stands for Reynolds number; υ stands for fluid kinematic viscosity ($m^2\cdot s^{-1}$); b stands for the characteristic length, which is seen as the chord length.

In the experiment, when the flow velocity is higher than a specific value, the lift coefficient (or the drag coefficient) remained unchanged, which is considered to be an automatic model area. The lift coefficient (or the drag coefficient) under a different angle of attack is defined as the average value in the automatic model area. The lift coefficient and drag coefficient discussed in this study are both the average value in the automatic model area.

3. Results and analysis

3.1. Automatic model area

As shown in figures 3-5, when the Reynolds number is higher than 2.08×10^5, the lift coefficient or the drag coefficient value remain stable. We considered the lift coefficient (or the drag coefficient) in an automatic model area. The average value of the lift coefficient at the Reynolds number between 2.08×10^5~2.60×10^5 was considered to be the lift coefficient for this angle of attack.

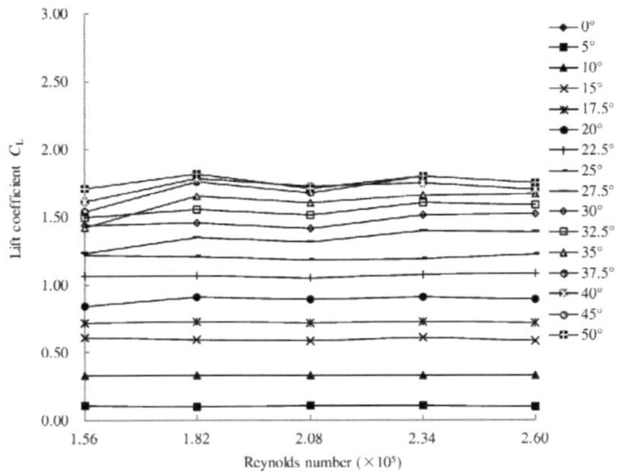

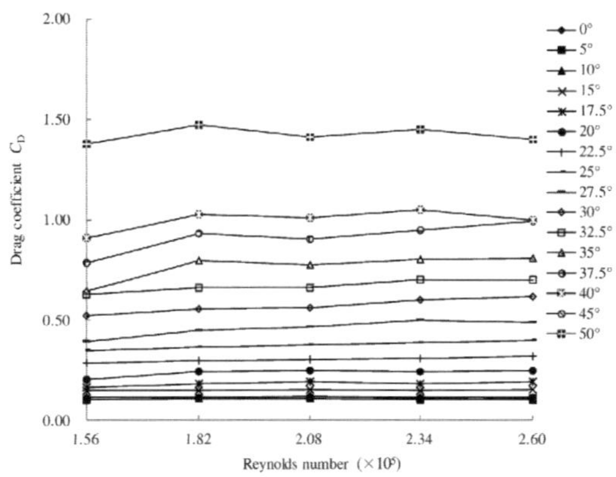

Figure 3. The relationship between lift coefficient, drag coefficient, and Reynolds number of the D1 model.

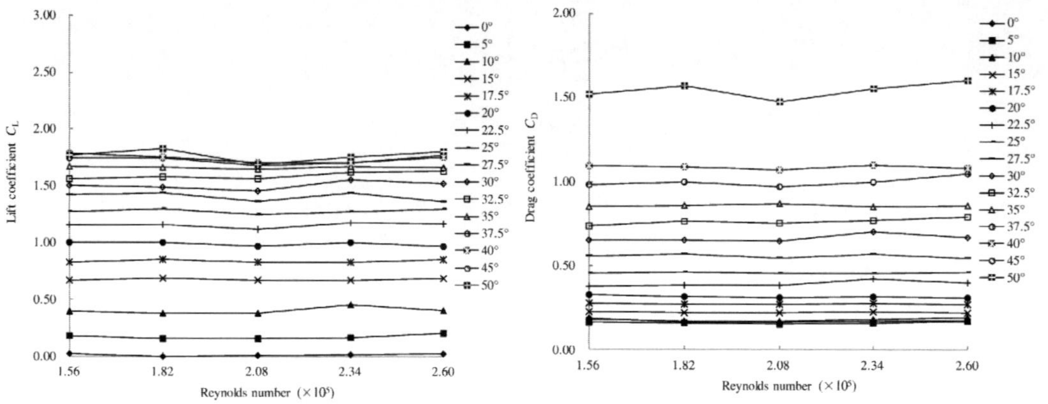

Figure 4. The relationship between lift coefficient, drag coefficient, and Reynolds number of the D2 model.

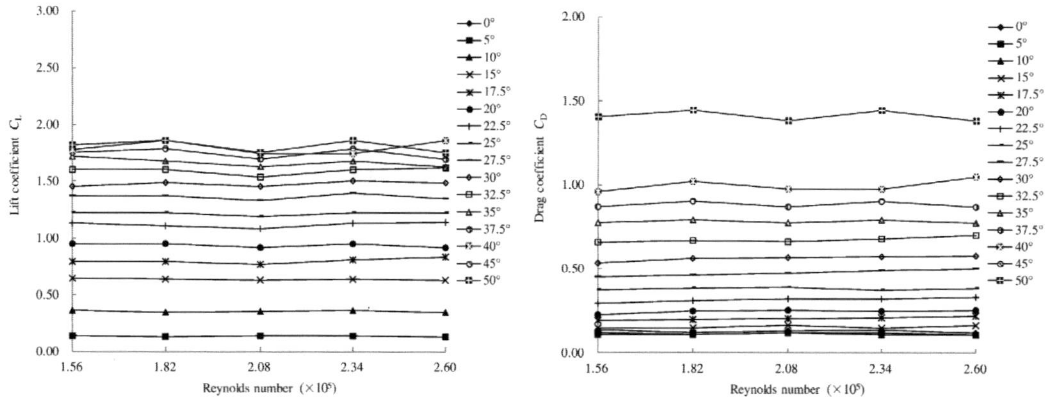

Figure 5. The relationship between lift coefficient, drag coefficient, and Reynolds number of the D3 model.

3.2. Hydrodynamic performance

The results of the lift coefficient, drag coefficient, and lift drag ratio under different angles of attack are shown in figure 6. The results of hydrodynamic performance of the otter board, show that the C_L and K value of the otter board show a trend of increasing at the beginning and then decreasing with the increase of the angle of attack. The C_D value reflects an upward trend with the increase in the angle of attack.

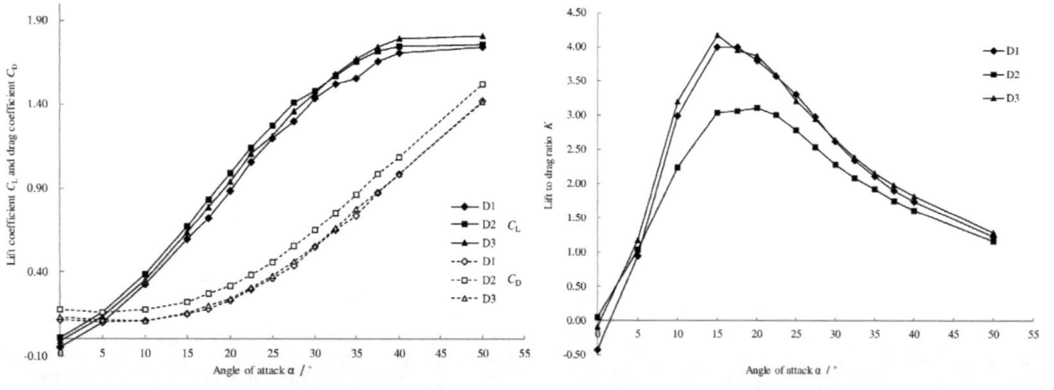

(a) (b)

Figure 6. Hydrodynamic performance of the otter board at different angles of attack.

For otter board D1, at the critical angle of attack α=50°, the max lift coefficient ($C_{L\,max}$) is 1.746. In this case C_D=1.422 and K=1.228. When the working scope of the angle of attack is between 15°~30°, C_L>0.597 and K>1.027. The mean value of the lift coefficient was 1.027 and the mean of lift to drag ratio was 3.464. When α=17.5°, the max lift to drag ratio (K_{max}) is 3.997, in this case C_L=0.720, C_D=0.180.

For otter board D2, at the critical angle of attack α=50°, $C_{L\,max}$=1.761, in this case C_D =1.521 and K=1.157. When the working scope of the angle of attack is between 15°~30°, C_L>0.671, and K>2.284. The mean value of the lift coefficient was 1.114 and the mean of lift to drag ratio was 2.827. When α=20°, the max lift to drag ratio (K_{max}) is 3.101, in this case C_L=0.988 and C_D =0.319.

For otter board D3, at the critical angle of attack α=50°, $C_{L\,max}$=1.809, in this case C_D =1.413 and K=1.281. When the working scope of the angle of attack is between 15°~30°, C_L>0.633, and K>2.643. The mean value of the lift coefficient was 1.071 and the mean of lift to drag ratio was 3.482. When α=15°, K_{max}=4.165, and in this case C_L=0.633, C_D =0.152.

When the scope of angle of attack is between 0°~30°, the C_L value shows a trend of first rising then declining with the increase of the angle of the front flow deflector. When the scope of the angle of attack is between 32.5°~50°, the C_L value shows an uptrend with the increase of the angle of the front flow deflector. For the D3 otter board, when the working scope of the angle of attack is between 15°~30°, the numerical range of C_L is 0.633~1.464 and the numerical range of K is 2.643~4.165. In this case, the lift coefficient value of D2 is higher than the D1 and D3 otter boards. However, the lift to drag ratio of the D2 otter board is far below the other otter boards because its drag coefficient is the highest. The lift coefficient of the D3 otter board is higher than the D1 otter board, which supported the conclusion that the D3 otter board has a better hydrodynamic performance, although its lift to drag ratio is slightly lower than the D1 otter board. For otter board D3, at the working angle of attack α=30°, C_{Lmax}=1.464, C_D =0.554, and K=2.643. When α=15°, K_{max}=4.165, C_L=0.633, and C_D =0.152. This suggests that the best working scope of the angle of attack is between 15°~30°, in which case, C_L>0.633 and K>2.643. The mean value of the lift coefficient was 1.071 and the mean of the lift to drag ratio was 3.482.

4. Discussion

The lift coefficient and lift to drag ratio of the otter board at different front flow deflector angles were compared and analyzed. The analysis showed that the D3 otter board had a better hydrodynamic performance than the other models. In the angle of attack between 15°~30°, the lift coefficient of the D3 otter board was higher than the D1 otter board, which supported the conclusion that the D3 otter board had a better hydrodynamic performance, although its lift to drag ratio was slightly lower than the D1 otter board. In this working scope of the angle of attack, the numerical range of C_L of the D3 otter board was 0.633~1.464 and the numerical range of K was 2.643~4.165.

The key factor in achieving highly efficient bottom trawl fishing is the reasonable selection of the otter board using performance and operation characteristics [1,3,5]. Table 3 shows comparisons of the hydrodynamic performance of several commonly used offshore trawl otter boards. The rectangular flat and oval flat slotted otter boards have a simple construction, and are less affected by working conditions. However, the expansion performance of the rectangular flat and oval flat slotted otter board is poor, and the maximum lift coefficient is less than 1.0 [2]. The hydrodynamic performance of the oval cambered single slotted otter board is better than the rectangular flat and oval flat slotted otter board [2,3]. The lift coefficient of the V type otter board is higher than the plat otter board, but the lift to drag ratio is lower than other types of otter boards, which means that the expansion efficiency is poor [6]. The lift characteristics of the vertical cambered otter board showed a large improvement over the flat otter board. The maximum lift to drag ratio of the vertical cambered otter board is 4.67, which is higher than the other types of flat otter boards [2]. However, with the increase in the aspect ratio, the stability performance of vertical cambered otter board was lower than in the other types of otter boards. After adding the leading-edge slot, the expansion performance is improved, but the expansion efficiency decreased as the drag efficient increased. The maximum lift to drag ratio of the oval cambered double slotted otter board (proposed in this study) and the vertical cambered otter board are over 4.0, which means the expansion efficiency is significantly higher than found in other types of otter boards. The maximum lift coefficient of the oval cambered double slotted otter board is 1.464, which is the maximum value in several commonly used types. The maximum lift to drag ratio at α_0 is also the highest value for several types, which suggests that the D3 otter board has good expansion performance and efficiency.

The critical angle of attack of the lift coefficient increases with the increase of the flow deflector angle. This is because when the main panel has reached the critical angle, the diversion wing has not yet reached the critical angle of attack, which provides a part of the lift [1]. The mechanism of the change of the flow deflector is discussed below. The proportion of the "funnel shape" structure of the inlet and outlet changed with the change of the flow of the deflector angle. Vortex flow at the back of the main plates drifts backward along the flow through, causing reduction of the otter board. When other structural parameters are constant, adjusting the flow deflector angle can optimize the hydrodynamic performance of the otter board. Recently, a design optimization approach for otter boards using computational fluid dynamics models showed that when the front flow deflector angle increased from 20° to 24°, the lift coefficient remained constant but the lift to drag ratio increased by 12% [22]. The configuration of the location of flow deflector is also an important factor affecting the hydrodynamic performance of the otter board, with the exception of the angle parameter. This study is limited because the lift force and drag force of the 3 parameter variations of the front flow deflector angle are measured. In future research, we can increase the testing of the gradient of the flow deflector angle. Analysis of the influence of deflector flaps structure parameters on the hydrodynamic performance of the otter board could be conducted using the computer simulation method, from the angle of the flow field effect.

Table 3. Hydrodynamic performance comparison among different types of otter boards.

Types	Working angle of attack α_0	Maximum lift coefficient C_{Lmax}	Maximum lift to drag ratio at α_0 $K_{\alpha 0}$	Maximum lift to drag ratio K_{max}
Rectangular flat [2]	40°	0.82	1.14	2.23
V type [6]	20°	1.08	0.85	1.86
Oval flat slotted [2]	35°	0.86	1.36	2.35
Oval cambered single slotted [2-3]	35°	0.93	1.25	2.10
Vertical cambered [2]	30°	1.44	2.21	4.67
This study D3	30°	1.46	2.64	4.17

5. Conclusion

Using the model experiment method, this study analyzed the hydrodynamic properties of a new type oval cambered double slotted otter board. Results show that the D3 otter board (with a front flow deflector angle of 29°) had a better hydrodynamic performance than the other tested boards. When α=15°, the max lift to drag ratio (K_{max}) was 4.165, C_L =0.633, and C_D =0.152. The experimental results show that the best working scope of angle of attack is between 15°~30°, in which case, C_L>0.633 and K>2.643. When α=30°, the max lift coefficient (C_{Lmax}) was 1.464, C_D =0.554, and K=2.643. The mean value of the lift coefficient was 1.071 and the mean of lift to drag ratio was 3.482. Comparative analysis of the hydrodynamic performance of different types of otter boards showed that the D3 otter board had both good expansion performance and good expansion efficiency.

Acknowledgments
This research was financially supported by the National Special Research Fund for Non-Profit sector (Agriculture) (201203018) and the National Natural Science Foundation of China (31402351).

References

[1] Guo G X, Liu T Y, Huang X H and Gu F L 2008 *Theory Research and Practice of the Hydrodynamic of Otter Board* (Guangzhou: Guangdong Science and Technology Press) pp 160-211
[2] Zhang X, Wang M Y and Xu B S 2004 A primary study on type, structure and performance of trawl otter board *J. Fish. Sci. China* **11** (suppl.) 107-13
[3] Zhuang X, Xing B B, Xu C C and Zhang G S 2015 Research overview on hydrodynamic characteristics of otter board *Fish. Mod.* **42** 63-8
[4] Broadhurst M K, Sterling D J and Cullis B R 2012 Effects of otter boards on catches of an Australian penaeid trawl *Fish. Res.* **131-133** 67-75
[5] Chen X Z and Huang X C 2011 *Theory and Method of Fishing Gear Model Test* (Shanghai: Shanghai Science and Technology Press) pp 388-425
[6] Li C C, Liang Z L, Huang L Y, Zhou W F, Sun P and Wang L 2013 Hydrodynamic study on a vee type otter board of small trawl vessel *Mar. Sci.* **37** 69-73
[7] Salaa A, Farranb J P, Antonijuan J and Lucchetti A 2009 Performance and impact on the seabed of an existing- and an experimental-otterboard Comparison between model testing and full-scale sea trials *Fish. Res.* **100** 156-66
[8] Takahashi Y, Fujimori Y, Hu F X, Shen S and Kimura N 2015 Design of trawl otter boards using computational fluid dynamics *Fish. Res.* **161** 400-7
[9] Park H H 2007 A method for estimating the gear shape of a mid-water trawl *Ocean Eng.* **34** 470-8.
[10] Prat J, Antonijuan J, Folch A, Salaa A and Lucchetti A 2008 A simplified model of the interaction of the trawl warps, the otter boards and netting drag *Fish. Res.* **94** 109-17
[11] Fukuda K, Hu F X, Tokai T and Matuda K 1999 Effects of aspect and camber ratios on hydrodynamic characteristics of biplane-type otter board *Jpn. Soc. Fish. Sci.* **65** 860-5
[12] Fukuda K, Hu F X, Tokai T and Matuda K 2000 Effect of aspect ratio on lift and drag coefficients of cambered plates *Jpn. Soc. Fish. Sci.* **66** 97-103
[13] Kinoshita H and Kumazawa T 2011 Taisei development of trawl gear with canvas kite *Jpn. Soc. Fish. Eng.* **48** 89-98
[14] Yamasaki S, Matsushita Y, Kawashima T, Tomiyama M and Kumazawa T 2007 Evaluation of a conventional otter board used in trawl fishery in Ise-wan Bay and proposal of a new design *Jpn. Soc. Fish. Sci.* **73** 220-5
[15] Zhang X, Wang J H, Wang M Y, Yu Y F and Xu B S 2004 Hydrodynamic characteristics of rectangular cambered V type otter board I: Relationship between cambered ratio of fairwater fin and hydrodynamic characteristics of otter board *J. Fish. Sci. China* **11** (suppl.) 5-8
[16] Wang J H, Zhang X, Wang M Y, Yu Y F and Xu B S 2004 Hydrodynamic characteristics of rectangular cambered V type otter board II: Effects of aspect ratio, slotted position and slotted width on hydrodynamic characteristics of otter board *J. Fish. Sci. China* **11** (suppl.) 9-13
[17] Wang L, Wang L M, Yu W W, Feng C L and Shi J G 2015 Influence of vane dimension on hydrodynamic performances of single slotted cambered otter board *Fish. Mod.* **42** 55-60
[18] Liu J, Huang H L, Chen S, Li L Z, Wu Y, Xu G D and Rao X 2013 Hydrodynamic characteristics of low aspect ratio vertical cambered otter board *J. Fish. China* **37** 1742-9
[19] Liu J, Huang H L, Chen S, Wu Y, Li L Z and Rao X 2014 Model test of the hydrodynamic

IOP Conf. Series: Earth and Environmental Science **77** (2017) 012012 doi:10.1088/1755-1315/77/1/012012

characteristics of two vertical cambered V type otter boards *Chinese J. Hydrodyn.* **29** 183-8

[20] Liu J, Huang H L, Wu Y, Rao X, Li L Z, Chen S, Yang J L and Zhou B 2015 Model test of hydrodynamic characteristics of two types of vertical cambered slotted otter boards *S. China Fish. Sci.* **11** 68-74

[21] Liu J, Huang H L, Chen Y, Wu X L, Wu Y, Chen S, Rao X and Li L Z 2015 Analysis on the hydrodynamic characteristics of Antarctic krill trawl otter board *Fish. Mod.* **42** 50-4

[22] Jonsson I M, Leifsson L, Koziel S, Tesfahunegn Y A and Bekasiewicz A 2015 Shape optimization of trawl-doors using variable-fidelity models and space mapping *Procedia Comput. Sci.* **51** 905-13

IOP Conf. Series: Earth and Environmental Science **77** (2017) 012013 doi:10.1088/1755-1315/77/1/012013

Mesh size selectivity of the gillnet in East China Sea

L Z Li[1,2], J H Tang[3], Y Xiong[3], H L Huang[1,2,4], L Wu[3], J J Shi[3], Y S Gao[3] and F Q Wu[3]

[1] Key Laboratory of East China Sea and Oceanic Fishery Resources Exploitation, Ministry of Agriculture, East China Sea Fisheries Research Institute, Chinese Academy of Fishery Sciences, Shanghai 200090, China
[2] Engineering Technology Research Center of Marine Fishing, Chinese Academy of Fishery Sciences, Shanghai 200090, China
[3] Marine and Fisheries Research Institute in Jiangsu Province, Nantong, 226007

E-mail: ecshhl@163.com

Abstract. A production test using several gillnets with various mesh sizes was carried out to discover the selectivity of gillnets in the East China Sea. The result showed that the composition of the catch species was synthetically affected by panel height and mesh size. The bycatch species of the 10-m nets were more than those of the 6-m nets. For target species, the effect of panel height on juvenile fish was ambiguous, but the number of juvenile fish declined quickly with the increase in mesh size. According to model deviance (*D*) and Akaike's information criterion, the bi-normal model provided the best fit for small yellow croaker (*Larimichthy polyactis*), and the relative retention was 0.2 and 1, respectively. For *Chelidonichthys spinosus*, the log-normal was the best model; the right tilt of the selectivity curve was obvious and well coincided with the original data. The contact population of small yellow croaker showed a bi-normal distribution, and body lengths ranged from 95 to 215 mm. The contact population of *C. spinosus* showed a normal distribution, and the body lengths ranged from 95 to 205 mm. These results can provide references for coastal fishery management.

1. Introduction

Gillnets have traditionally been used in offshore fisheries. The target species include small yellow croaker (*Larimichthy polyactis*), butterfish (*Pampus argenteus*) [1], spotted mackerel (*Scomberomorus niphonius*) [2], and so on. The annual output of gillnet is lower than trawl products only, and accounts for 20% of the total catches in offshore fishery. Gillnets are highly selective for those species captured mainly by gilling (i.e., captured behind the gill-cover) or wedged (being held by a mesh around their maximum body girth). Therefore, retention is supposed to increase with size up to a length of the maximum catch and decrease thereafter, and consequently the range of size-at-catch of a target species can be controlled with a careful choice of the mesh size [3]. It is a common method adopted by the management. The minimum mesh size regulation proposed the minimum mesh size (MMS) of gillnet in the offshore fishery to be 50 mm, but the selectivity of the 50-mm mesh gillnet has not been reported in the East China Sea and whether the regulation is reasonable was unknown. The limitation of MMS 50 mm mainly aimed for the gillnet targeted at small yellow croaker.

Small yellow croaker is an important economic species in coast fishery [4]. The annual catches have reached up to 10,000 tons in recent years, and are targeted by gillnet, trawl, stow net, and so on

[5]. According to the fishery resources study, the fishery resources of the small yellow croaker have declined seriously, and the population structure is simple. The catch weight of the first age accounts for 90%. Early sexual maturity and individual miniaturization are obvious [6]. Most researches attributed the fishery resources decline to the higher fishing intensity and environmental pollution. The government has adopted a series of policies, including prohibited fishing in summer, the MMS regulation for the fishing gear of marine fishery, and so on, to promote the sustainable development of the coast fishery.

Gillnet selectivity in the East China Sea (ECS) has been explored earlier [1, 2, 7, 8]. Some selectivity experiments on the gillnets for small yellow croaker were carried out in Korean waters [9, 10]. The optimal mesh size for 50% retention of yellow croaker (total length 191 mm) was estimated as 51.1 mm. The setting depth of the net, which is 4.6–13.8 m above the sea bottom, is the best to reduce bycatch and catch more large-size fishes [11]. However, no selectivity studies on the gillnet for small yellow croaker have been conducted in the East China Sea, and most of the size selectivity experiments have been carried out in trawl and stow net.

2. Materials and methods

2.1. Survey area and gears

The production tests were carried out from October 22 to November 5, 2011. The survey area was located in the traditional fishing ground in the north of East China Sea, 32°02′ – 33°16′N, 123°25′E-123°54 E′ (figure 1). The water depth was about 60 m; 73 valid hauls were finished.

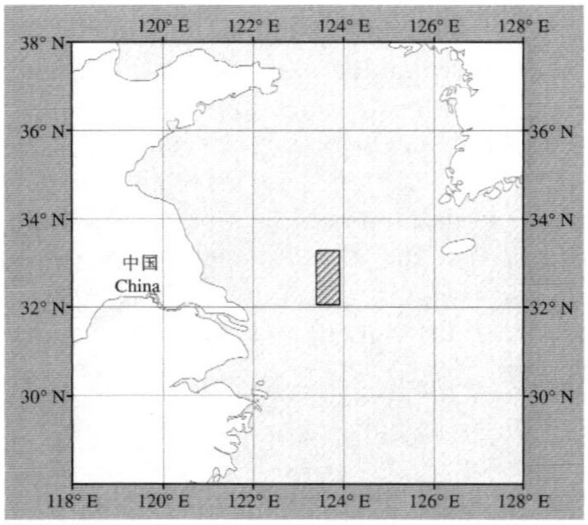

Figure 1. Survey area of the selectivity experiment.

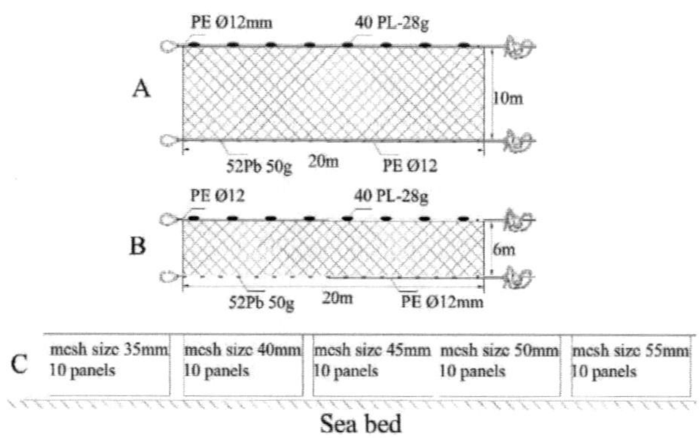

Figure 2. Configuration and arrangement of the test nets. The panel height of test net A was 10 m and that of test net B was 6 m. C is the arrangement of the test nets with different mesh sizes.

A total of 10 different gillnets were used, including 2 heights (6- and 10-m panels) and 5 mesh sizes (35 mm, 40 mm, 45 mm, 50 mm, and 55 mm) in each height. Ten panels of each kind test net were used in the production trial. The test net structure and arrangement are shown in figure 2. The foot and upper ropes of each panel were 20 m long. The horizontal hang ratio was 0.6. The nets were made of nylon monofilament with diameter 0.2 mm. Gillnets sewn together were set on the bottom in appropriate depth zones. The time of net deployment depended on weather conditions and water temperature. The soak duration was between 120 and 300 min. The configuration and arrangement of the test nets are shown in figure 2.

2.2. Catch handling

After hauling, all species were sorted out and identified to the species level. The body length was measured to the nearest millimeter and reduced to length frequency data, which was used for mesh size selectivity analysis. Sex was identified by anatomy. The catch data in different trials were accumulated by the same test net.

2.3. Selectivity estimation

Some assumptions were commonly adopted to simplify the model parameter estimation; the equal fishing efforts of all test nets and the selection curves of the gillnets followed the "principle of geometric similarity" [12]. According to the assumption, the catch process was only dependent on the relative geometry of the fish and the mesh size.

The selection method used all the advantageous properties of the maximum likelihood and allowed for the incorporation of between-haul variability [3, 13], further satisfying the uni-modal [14] and bi-normal selection curve parameter estimation [15] used herein.

According to the selection method, for a given length class j, the numbers of fish N_{ij},which encountered the test net i, were assumed to be the independent possion random variables, and N_{ij} could be described as the product of the abundance of length class j fish (N_j) and the relative fishing intensity of gillnet I (q_i) [14]. The catch numbers of the length class j in gillnet i (C_{ij}) was distributed as follows:

$$C_{ij} \sim Po(q_i N_j S_{ij}) \quad (1)$$

S_{ij} was the retention probability of gillnet i to length class j. The log-linear form was as follows:

$$lnP = \sum_i \sum_j [C_{ij} \ln(\lambda) - lnC_{ij}! - \lambda] \quad (2)$$

C_{ij} was the number of catches. $\lambda = qN_j S_{ij}$ was the expected value if the selectivity curve was known. The selectivity curve parameters and the relative abundance of the given length class j could be estimated simultaneously.

The commonly used selectivity curves (uni-normal and bi-normal models) [14, 16, 17] are shown in table 1, which were also used in this paper.

Table 1. Commonly used selectivity model.

Model	Selection curve	Note
Normal	$\exp\left(-\frac{(\frac{l_j}{m_i} - R_0)^2}{2\sigma^2}\right)$	l_j is the body length of the j length group; m_i is the mesh size of the i kind test nets
Gamma	$\left[\frac{l_j}{(\alpha-1)\beta}\right]^{\alpha-1} \exp\left(\alpha - 1 - \frac{l_j}{\beta}\right)$	R_0 ; R_{01} ; R_{02} is the selectivity factor
Log-normal	$\exp\left\{-\frac{\left[\ln(\frac{l_j}{m_i}) - \ln(R_0)\right]^2}{2\sigma^2}\right\}$	σ; σ_1 ; σ_2 is the model standard error; α and β are model parameters; ω is weight coefficient
Bi-normal	$\exp\left[-\frac{(\frac{l_j}{m_i} - R_{01})^2}{2\sigma_1^2}\right] + \omega \exp\left[-\frac{(\frac{l_j}{m_i} - R_{02})^2}{2\sigma_2^2}\right]$	

3. Results

3.1. Effect of panel height and mesh size on catches

IOP Conf. Series: Earth and Environmental Science **77** (2017) 012013 doi:10.1088/1755-1315/77/1/012013

The species composition of the catch was affected synthetically by panel height and mesh size. Table 2 presents the composition of species that were captured by different panels. Altogether, 21 species were captured by the 6-m panels and 30 species by the 10-m panels. With the 6-m panels, the species number declined along with the increasing mesh size. However, with the 10-m panels, a peak number of the species were captured by the 45-mm mesh.

Table 2. Species composition and catch number by different test nets.

Species	Panel height 6 m					Panel height 10 m				
	35 mm	40 mm	45 mm	50 mm	55 mm	35 mm	40 mm	45 mm	50 mm	55 mm
Larimichthy polyactis	155	158	67	53	38	62	164	84	76	43
Chelidonichthys spinosus	10	14	18	113	48	29	27	124	64	65
Pneumatophorus japonicus	23	22	23	18	11	16	9	22	24	15
Engraulis japonicas Temminek	3	4		1	16	11	17	18	8	9
Johnius belengerii	14	7	3	3	4	15	26	3	6	3
Setipinna taty	10	7	2	4	3	7	17	2	2	
Decapterus maruadsi	16	4	4		2	2	16	1	3	3
Chaeturichthys stigmatias Richardson	1	2	1	1	1	6	6	10	7	6
Pennahia argentatus	1	2	3	9	2	1		4	6	10
Inimicus japonicus	4	7			1	2	5	4	1	3
Liparis tanakae			1	2	5			1	1	1
Cynoglossus joyneri Günther	2			3			2	2	1	
Scomberomorus niphonius				1	2		1		4	1
Lophius litulon				1	1				2	2
Caelorhynchus multispinulosus Katayama		1				2	1			
Trichiurus japonicus								2	1	
Psenopsis anomala						1				1
Cleisthenes herzensteini								2		
Miichthys miiuy	1							1		
Clupea pallasi Valencienes		2								
Saurida elongata									1	
Pleuronectes platessa Linnaeus								1		
Nibea albiflora										1
Zoares gilli Jordan				1						
Collichthys lucidus									1	
Parapercis sxefasciata		1								
Gnathagnus elongatus										1
Todarodes pacificus Steenstrup									2	
Trachypenaeus curvirostris	3	1	1					2		
Parapenaeus fissuroides Crosnier								1		
Portunus trituberculatus								1		
Charybdis feriatus	1							2		
Oratosquilla nepa									1	

The dominant species caught by the 6- and 10-m panels were the same. For the 6-m panels, 351 small yellow croaker and 203 *C. spinosus* accounted for 71.47% of total catches. For the 10-m panel, 327 small yellow croaker and 309 *C. spinosus* accounted for 66.67% of total catches. The body-length range of the dominant species was also similar between the 6- and 10-m panels. The majority of the small yellow croaker fish were concentrated within the range of 88–128 mm and 138–178 mm, and *C. spinosus* was concentrated within the range of 118–193 mm.

Table 3 presents the proportion and the catch number of small yellow croaker juvenile fish for different panels. Although the juvenile fish number and the proportion of the 10-m panels were slightly larger than those of the 6-m panels, the effect of panel height on juvenile fish composition was ambiguous for any single mesh size.

Table 3. Small yellow croaker juvenile fish proportion and number captured by different test nets.

Panel height		Juvenile fish number or proportion of different mesh sizes					Total proportion of juvenile fish
		35 mm	40 mm	45 mm	50 mm	55 mm	
6 m	Number	72	33	17	22	18	162
	Proportion	63.16%	37.08%	25.37%	47.83%	51.42%	46.15%
10 m	Number	38	67	32	22	20	179
	Proportion	71.70%	59.82%	48.48%	39.29%	50%	54.74%

Note: According to the allowable size of capture and juvenile proportion of key marine fishery resources of Zhejiang province, the minimum landing size of small yellow croaker is 145 mm.

C. spinosus was not an economic species, and the basic biological research was limited. No minimum allowed landing size (MALS) was currently reported; therefore, the proportion of juvenile fish was not calculated in the present study.

Along with the increasing mesh size, the number of juvenile fish declined quickly; however, the proportion first declined and then increased. This difference might be related to the lack of larger fish. The most efficient mesh size for small yellow croaker was 40 mm. The number of fish below the MALS notably decreased along with the increasing mesh size. The most efficient mesh size for *C. spinosus* was 50 mm, and 117 fishes were captured. The catches declined with changes in mesh size.

3.2. Selectivity estimation

The body-length range of the dominant species was similar between the 6- and 10-m panels, and the catches of single-height panels were not enough for selectivity analysis. Therefore, when the mesh size of 6- and 10-m panels was the same, the catch data were accumulated for the mesh size selectivity analysis. The length frequency distribution for the species analyzed is shown in figure 3 (dotted line). Small yellow croaker and *C. spinosus* were the only species for which sufficient data were available for selection curve estimation.

The length frequencies of small yellow croaker were distributed with a similar mode for different mesh sizes. Two visual peaks existed for each mesh size. Along with increasing mesh size, the length related to the first peak declined. However, the second peak showed the opposite response. For *C. spinosus*, the length frequency distribution was uni-modal, increased faster on the left side of the peak, and gradually declined on the right side.

Based on the length frequency distribution data, four types of selectivity models were used to fit the data. Table 4 presents the fitting results (three uni-models and one bi-normal model). According to the Akaike's information criterion (AIC) and model deviance (D), for small yellow croaker, the bi-normal model provided the best fit, the AIC was the smallest, and $P > 0.01$. For *C. spinosus*, the log-normal was the best model, the AIC was the smallest, and $P > 0.01$.

Together with MALS (145 mm), the selection curve was plotted in figure 3 for small yellow croaker. Two separate peaks were present in the fitted bi-normal curves, and the relative retention was 0.2 and 1, respectively. This figure demonstrated that, except for the 50- and 55-mm mesh, the retention rates of the MALS were relatively higher for other mesh sizes.

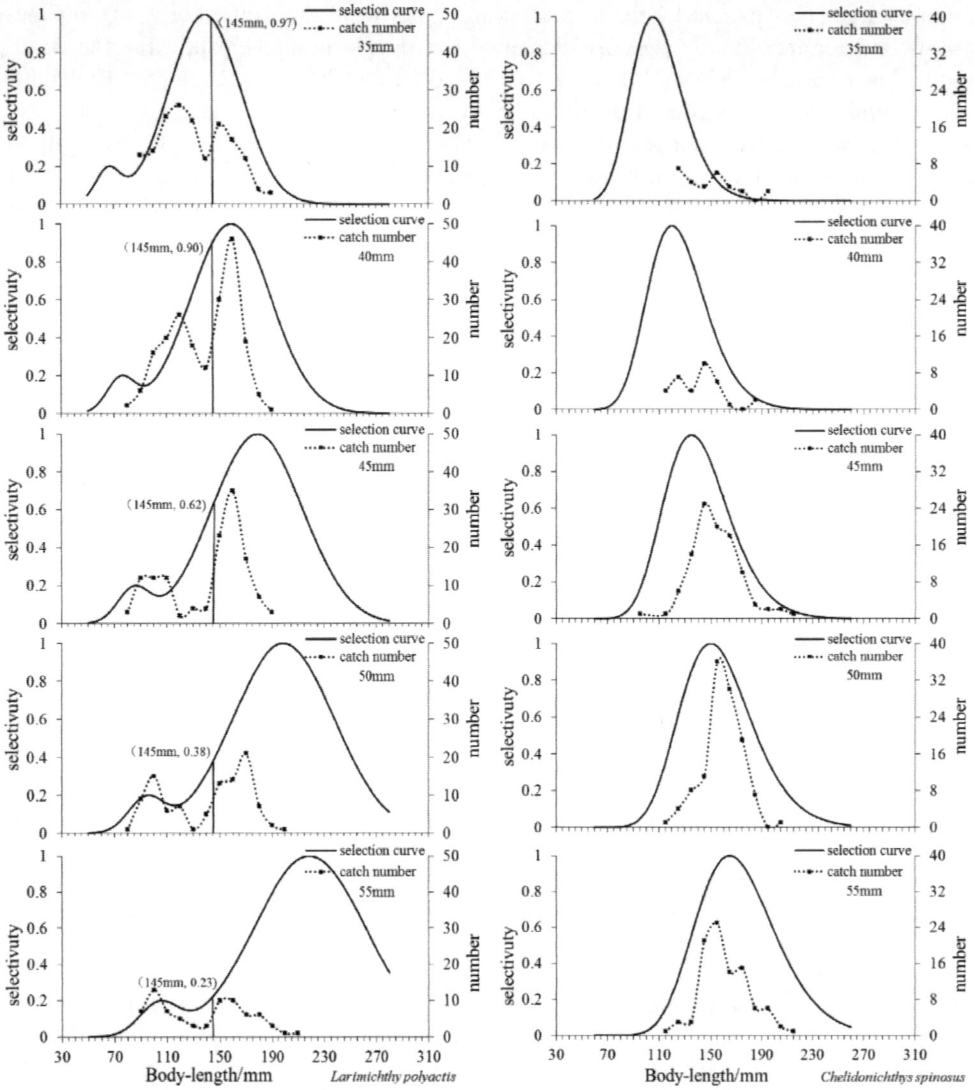

Figure 3. Body-length frequency distribution of *Larimichthy polyactis* and *Chelidonichthys spinosus* captured by test nets (dotted line) and the mesh size selection curve (solid line).

Table 4. Fitting result of different selectivity models.

Species	Model	Parameters	Model devian ce	df
Larimichthy polyactis	Normal	$(R_0, \alpha) = (-4.17, 1.17)$	114.76	54
	Gamma	$(\alpha, \beta) = ((-6.69, 0.83)$	129.80	54
	Log-normal	$(R_0, \alpha) = ((-5.83, 0.59)$	134.40	54
	Bi-normal	$(R_{01}, R_{02}, \sigma_1, \sigma_1, \omega) = (3.96, 1.86, 0.61, 0.05, 0.19)$	71.16	51
Chelidonichthys spinosus	Normal	$(R_0, \alpha) = (2.88, 0.71)$	68.75	46
	Gamma	$(\alpha, \beta) = (-19.31, 0.16)$	88.04	46
	Log-normal	$(R_0, \alpha) = (3.00, 0.03)$	51.89	46
	Bi-normal	$(R_{01}, R_{02}, \sigma_1, \sigma_1, \omega) = (2.88, 2, 0.71, 0.50, 0. 70)$	65.30	43

For *C. spinosus*, the right tilt of the log-normal curve was obvious. The selectivity increased faster on the left side of the model peak, with a gradual decline on the right side; these coincided well with the original data.

3.3. Estimated relative abundance of fish contacting the nets

According to the best selectivity model, the contact population of small yellow croaker showed bi-normal distributions, and body length ranged from 95 to 215 mm (figure 4). There were two frequency peaks at 100 and 160 mm body length, respectively, indicating two age groups. The biomass of juvenile fish accounted for 72.3%.

Figure 5 shows that the contact population of *Chelidonichthys spinosus* was normally distributed, and body length ranged from 95 to 205 mm. The frequency peak was at 155 mm. The right tilt of the abundance plot was obvious.

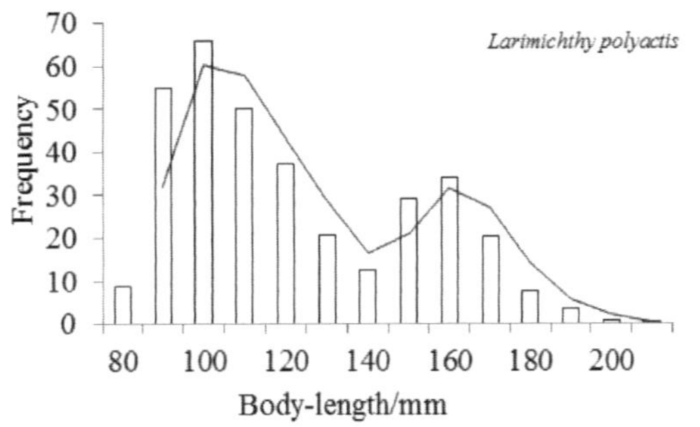

Figure 4. Estimated population structure of small yellow croaker.

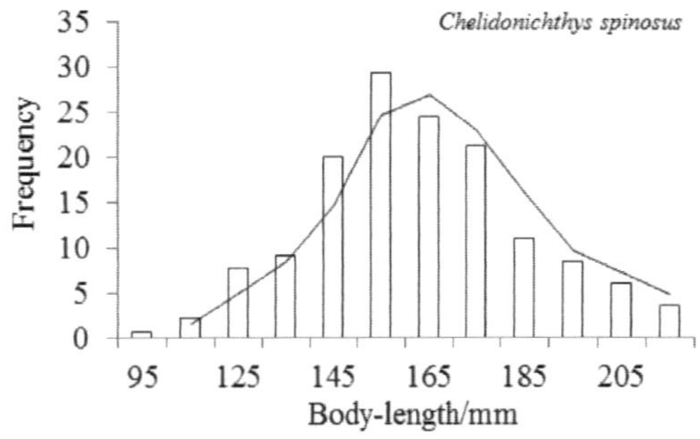

Figure 5. Estimated population structure of *Chelidonichthys spinosus.*

4. Discussion

Previous studies on small yellow croaker in Korean waters showed [7,8] that a bi-normal model also provided the best fit; the relative total length at a maximum selectivity of the primary fishing modes was estimated to be 4.03, which was similar to that in the present study. However, the relative total length at a maximum selectivity of the secondary fishing modes was larger than the results of the present study and was estimated to be 4.62. The bi-normal model likely reflected the combined effects of several capture processes, where the primary mode would result from gilling or wedging and the secondary mode from other methods of capture. The majority of small yellow croaker were concentrated within the range of 88–128 mm for the secondary fishing modes in the present study. The size was significantly smaller than that of the catches in previous studies. Therefore, it was suggested that the capture modes were different for the secondary fishing modes and led to the estimation variance. The optimal mesh size for 50% retention for small yellow croaker (total length of 191 mm) was estimated as 51.1 mm. According to the selectivity curve used in the present study, the body length for 50% retention ($L_{0.5}$) of single drift nets with a mesh size of 51.1 mm was 156.5 mm and transfer to total length was 186.4 mm. Therefore, the two results were similar. An MMS of 50 mm was allowed in coastal gillnet fisheries. The $L_{0.5}$ of the 50-mm mesh size for small yellow croaker was 152 mm, which was larger than the minimum allowed landing size [18]. The regulation could satisfy fishery resources conservation.

Small yellow croaker is one of the main economic species targeted by trawling, stow nets, and gillnets in coastal fisheries. The MMS allowed for trawls in the East China Sea is 54 mm. According to studies in recent years, the $L_{0.5}$ of small yellow croaker caught by trawls with the MMS 54 mm was 113~130 mm [19, 20] and significantly less than the $L_{0.5}$ of small yellow croaker caught by gillnets with the similar mesh size. It could be concluded that, with similar MMS, gillnets were much more selective than trawls and stow nets.

Small yellow croaker is demersal fish. Along with the growth of juvenile fish, the habitat changed from demersal to semi-demersal water or pelagic water [21]. Therefore, the set depth of gillnet or the panel height was relative to catch composition. Taeg [11] concluded that a setting depth of 4.6–13.8 m above the sea floor was the best to reduce bycatch and catch larger fish in Korea coastal water. The lower the setting depth above the sea floor, the more the bycatch species captured. In this study, although the number and proportion of small yellow croaker juvenile fish for the 10-m panels were greater than those for the 6-m panels, the difference was not obvious. Further, the dominant species was the same, and the bycatch species increased with the filtering area of the nets; 22 and 30 species were captured by the 6- and 10-m panel, respectively.

Gillnetting is less limited by external factors (e.g., rough bottom) and has been widely used for fishery resource investigation [22-24]. However, according to the high selectivity of gillnets for target species, the gillnetting as a sampling method has been debated. The gillnet selectivity is usually defined as a product of the contact ratio and retention ratio. The probability of a fish to encounter and retain in a net increases with swimming speed, body size, and so on. Therefore, the bigger fish will be captured more effectively, and the biomass may be overestimated [22-24]. The population structure of fish contacting nets was estimated in this study based on the better selectivity model; the contact population of small yellow croaker showed bi-normal distributions, and the juvenile fishes accounted for 72.3%. This value was significantly smaller than that in the previous study based on trawling, and the population structure was normal distribution [5, 18]. It is doubtless that trawling cannot provide a nonbiased result for fishery resource estimation. Trawl catchability can be relatively low for large fishes, avoiding the trawl but not gillnets [24]. Hence, a good strategy is to use several sampling methods and cover the shortage [23]. For juvenile fish, the biomass estimation from gillnetting can be used as the lower limit, and the result from trawling can be used as the upper limit. Therefore, it can be concluded that the proportion of small yellow croaker juvenile fish should be more than 72.3%. The population structure is simple, and the stock in Chinese coastal waters has been overexploited. For *C. spinosus*, the population structure distribution is normal and the right tilt is obvious.

As the basic biological research on *C. spinosus* is limited, it is not possible to estimate the influence of the test nets on the resource population of this species. The selectivity analysis can provide a reference for future research.

Acknowledgments
This work was supported by the Special Fund for Agro-Scientific Research in the Public Interest of China (201203018).

References
[1] Li L Z, Huang H L, Wang L, Feng C L, Xu Y J and Zhang Y 2010 The mesh size selectivity research of silvery pomfret gillnet in Yellow and East China Sea *Marine Fish* **32(1)** 89-94
[2] You Z B, Li X S, Zhao X Y, Wang X L, Sun Z Z and Zhu J C 2014 The selectivity research of big mesh gillnet for *Scomberomorus niphonius J. Fish. China* **38(2)** 289-96
[3] Paulo F, Rogélia M, Aida C and Preciosa S 2005 Gill-net selectivity off the Portuguese western coast *Fish. Res.* **73** 323-39
[4] Zhang Q H, Cheng J H and Xu H X 2007 *Biological Resources and the Environment in the East China Sea Continental Shelf* (Shanghai: Fu Dan University Press) pp 472-87
[5] Liu Z L, Yan L P, Yuan X W and Ying W 2013 Stock assessment of small yellow croaker in the East China Sea based on multi-source data *J. Fish. Sci. China* **20(5)** 1039-49
[6] Liu Z L, Xie H Y, Yan L P,Yuan X W, Yang L L, Li Y X, Wu Y and Chen J H 2013 Comparative population dynamics of small yellow croaker *Larimichthys polyactis* in Southern Yellow Sea and East China Sea *J. Dalian. Oce. Uni.* **28(6)** 627-32
[7] Zhang J and Sun M C 2006 Research progress on size selectivity of gillnet *J. Fish. Sci. China* **13(6)** 1040-8
[8] Zhang P, Yang L, Zhang X F and Tang Y G 2005 Study on selectivity of different mesh size

gillnet for *nemipterus virgatus* in South China Sea *South China Fish. Sci.* **11(12)** 61-6
[9] Kim S H, Park S W, Bae J H and Kim Y H 2009 Mesh selectivity of drift gill net for yellow croaker, larimichthy polyactis, in the Coastal Sea of Gageo-do *J. Kor. Soc. Fish. Tech.* **42(5)** 518-22
[10] Kim S H, Park S W, Lee K H and Yang Y S 2012 The estimation of the optimum mesh size selectivity of a drift net for yellow croaker (Larimichthy polyactis) using by the SELECT model *J. Kor. Soc. Fish. Tech.* **48(1)** 10-9
[11] Oh T Y, Cho Y B, Seo Y I, Kim B Y and Lee C H 2014 Fishing characteristics of small yellow croaker drift gill net by net height *J. Kor. Soc. Fish. Tech.* **50(3)** 368-77
[12] Baranov F I 1948 *Theory and Assessment of Fishing Gear: Theory of Fishing with Gill-nets* chapter 7 (Moscow: Fish Industry Press) (Translated from Russian by the Ontario Department of Lands and Forest, Maple, Ont.).
[13] Millar R B 2000 Untangling the confusion surrounding the estimation of gillnet selectivity *Can. J. Fish Aquat. Sci.* **57** 507-11
[14] Millar R B and Hoist R 1997 Estimation of gillnet and hook selectivity using log-linear models *ICES J. Mar. Sci.* **54** 471-7
[15] Madsen N, Houst R, Wileman D and Moth-poulsen 1999 Size selectivity of sole gill nets fished in the North Sea *Fish. Res.* **44** 59-73
[16] Hovgård H 1996 A two-step approach to estimating selectivity and fishing power of research gill-nets used in Greenland waters *Can. J. fish. Aquat. Sci.* **53** 1007-13
[17] Sun M C 2004 *Fishing gear selectivity* (Beijing: Chinese Agriculture Press) pp 109-160
[18] Lin L S, Cheng J H, Lin J Z and Zhang H Y 2006 First capture sizes of major commercial fishes in the East China Sea Region *J. Fish. Sci. China* **13(2)** 250-6
[19] Huang H L, Wang M Y, Xu B S, Zhang X and Tang Z M 2005 Study on selectivity of mesh size of cod-end of trawl in the East China Sea region *J. Fish. China* **29(2)** 232-7
[20] Song X F, Chen X Z, Huang H L, Tang F H, Wang D H and Qu T C 2015 Selectivity of *Parimichthys polyactis* of bottom trawl in the East China Sea *J. Shanghai Oce. Uni.* **24(3)** 449-55
[21] Xu B D and Jin X S 2005 Variations in fish community structure during winter in the southern Yellow Sea over the period 1985-2002 *Fish. Res.* **71** 79-91
[22] Reidar B, Morten H, Kjell E M and Martin A S 2015 Recorded bimodal length frequency distributions of Arctic charr,Salvelinus alpinus (L.), and brown trout, Salmo trutta L.: An effect of both population structure and sampling bias *Polar biol.* **38** 895-903
[23] Juha J, Irma K and Ari L 2011 Quality and status of fish stocks in lakes: Gillnetting, seining, trawling and hydroacoustics as sampling methods *Hydrobiologia* **660** 29-36
[24] Olin M and Malinen T 2003 Comparison of gillnet and trawl in diurnal fish community sampling *Hydrobiologia* **506-509** 443-9

IOP Conf. Series: Earth and Environmental Science **77** (2017) 012014 doi:10.1088/1755-1315/77/1/012014

The changes of oil palm roots cell wall lipids during pathogenesis of *Ganoderma boninense*

A Alexander[1], J Dayou[2], S Abdullah[1] and K P Chong[1,3,4]

[1] Biotechnology Programme, Faculty of Science and Natural Resources, Universiti Malaysia Sabah, 88400, Kota Kinabalu, Sabah, Malaysia
[2] Energy, Vibration and Sound Research Group, Faculty of Science and Natural Resources, Universiti Malaysia Sabah, Jalan UMS, 88400 Kota Kinabalu, Sabah, Malaysia
[3] FGV Chair on Sustainable Oil Palm Management, Faculty of Agriculture, Universiti Malaysia Sabah, Locked Bag No. 3, 90509, Sandakan, Sabah, Malaysia

E-mail: chongkp@ums.edu.my

Abstract. One of the first physical defences of plants against fungal infection is their cell wall. Interaction between combinations of metabolism enzymes known as the "weapons" of pathogen and the host cell wall probably determines the fate of possible invasion of the pathogen in the host. The present work aims to study the biochemical changes of cell wall lipids of oil palm roots and to determine novel information on root cell wall composition during pathogenesis of *Ganoderma boninense* by using Gas Chromatography Mass Spectrometry. Based on Total Ion Chromatogram analysis, 67 compounds were found more abundant in the roots infected with *G. boninense* compared to the healthy roots (60 compounds). Interestingly, nine new compounds were identified from the cell wall lipids of roots infected with *G. boninense*. These includes Cyclohexane, 1,2-dimethyl-, Methyl 2-hydroxy 16-methyl-heptadecanoate, 2-Propenoic acid, methyl ester, Methyl 9-oxohexacosanoate, 5-[(3,7,11,15-Tetramethylhexadecyl)oxy]thiophene-2carboxylic acid, Ergosta-5,7,22,24(28)-tetraen-3beta-ol, 7-Hydroxy-3',4'-methylenedioxyflavan, Glycine and (S)-4'-Hydroxy-4-methoxydalbergione, this may involve as response to pathogen invasion. This paper provides an original comparative lipidomic analysis of oil palm roots cell wall lipids in plant defence during pathogenesis of *G. boninense*.

1. Introduction

Ganoderma boninense is a white rot fungus, which is responsible for destructive basal stem rot (BSR) disease in oil palm. This is a major problem in oil palm plantations in Southeast Asia, especially in Malaysia and Indonesia. The economic loss due to this disease is estimated up to 500 million USD per year in Malaysia [1, 2]. Effects of BSR disease either occur from death of palms or yield reduction from standing palms [3]. Various detection and control measures have been developed in managing this disease but none of them is satisfactory. Lack of understanding on the biology of pathogen and host-pathogen interaction during pathogenesis contributes to poor management of basal stem rot (BSR) disease.

Different approaches on metabolic and molecular processes in oil palm against *G. boninense* have been evaluated to further understand the pathogen-host relationship. However, no research reported the changes in cell wall composition of oil palm during the interaction, despite, knowing its importance as the foremost interface during the interaction event.

The plant cell wall is the first lines of defensive barrier in plant immune response against infection, where initially all plant pathogens have to get-through in order to colonize the host. In order to penetrate this barrier, pathogenic fungi evolved arrays of wall degrading enzymes virulence factor for disease development. After the penetration, the cell wall becomes a battleground between the host and its pathogen. Cell wall is a dynamic reservoir of antimicrobial compounds and generates signaling molecules that trigger defense responses in the host [4, 5]. During plant-pathogen interaction, pathogens attempt to utilize the metabolism of host plants to suppress plant defense and to obtain nutrients for survival [6, 7]. At the same time, the plant synthesizes metabolites that serve as signals, sedatives or toxins to limit the proliferation of the pathogen [8]. The importance of plant cell wall integrity and cell wall-mediated resistance during plant-pathogen interaction has been well demonstrated, but the related components have not been fully elucidated [7].

Lately, lipids were reported to trigger multiple stages of plant-pathogen interaction which includes communication between the host pathogen, activation and implementation of plant defenses in response to pathogen attack [9, 10]. Previous study [11] suggested lipids metabolites involved in early defense mechanism of oil palm against *G. boninense.* Therefore, lipidomics has gained a lot of attention and became an emerging field of study. Thus, the aim of this study is to provide an insight of oil palm *Ganoderma*-BSR interaction at lipidomics level and to discover potential metabolites for early disease detection. Moreover, this information could also be harnessed in managing BSR disease in future.

2. Materials and methods

2.1. Plant material

Oil palm seedlings D x P (*Dura x pisifera*) of eight months old were purchased from Sawit Kinabalu Sdn. Bhd. and artificially inoculated with *G. boninense* [12]. The stock culture of *G. boninense* was obtained from Universiti Malaysia Sabah and identified prior before inoculation [13]. The seedlings were maintained in a nursery with 70% of shade protection. Control seedlings (healthy/uninoculated) and inoculated seedlings were watered and fertilized according to standard agronomic practices. The seedlings were uprooted after three months and washed under running tap water to remove all unwanted materials and air dried. The success of *Ganoderma* infection was re-confirmed by using *Ganoderma* Selective Media (GSM) [14] and through ergosterol analysis [15]. The dried roots were lyophilized for 24 hours and homogenized into fine powder by using a blender.

2.2. Isolation of cell wall

The cell wall composition of control and infected roots were extracted according to procedures described by [16] with slight modification. All procedures were conducted at 4°C unless mentioned otherwise. Three gram of powdered oil palm root tissues were agitated overnight in 50 mL of pre-cooled extraction buffer (5 mM acetate, 0.4 M sucrose, pH 4.6) added with 1% of Polyvinylpolypyrrolidone using an orbital shaker at 180rpm. Cell walls were separated from soluble cytoplasmic fluid by centrifuging at 1,000 xg for 15 min. The pellet was resuspended in 50 mL of 5 mM acetate buffer at pH 4.6 with 0.6 M sucrose. The slurry was centrifuged at 1,000 xg for 15 min. The pellet was further washed in 50 mL of 5 mM acetate buffer at pH 4.6 with 1 M sucrose and centrifuged at 1,000 xg for 15 min. The residue was washed extensively with 50 mL of 5 mM acetate buffer at pH 4.6 and centrifuged at 1,000 xg for 15 min, thrice. The supernatant was discarded and final pellet was lyophilized overnight. Most intercellular components were removed from the cell wall given the advantage of sucrose gradients and extensive washing with low ionic strength of acidic buffer.

2.3. Cell wall lipids extraction

A small-scale modified protocol was adopted from [17] using 50 mL reaction tubes with screw-cap lids for extraction of cell wall lipids. Three replicates of one gram of each control and infected samples (six replicates) were added to the reaction tubes along with 20 mL of solvent mixture (Chloroform: Methanol; 2:1; v/v). The tubes were vortexed and agitated on an orbital shaker for one hour at room temperature. The tubes were then centrifuged at 2000 rpm for five min to recover the liquid phases. The solvent mixture were washed with 0.2 volume of miliQ® water and vortexed for few seconds. The tubes were centrifuged again at 2000 rpm to separate the mixture into two phases. The aqueous upper phases were removed and the lower chloroform phases containing lipids were kept for further use. The extracts were later dried under a stream of nitrogen gaseous at room temperature.

2.4. Mild methylation of fatty acid methyl esters (FAMEs)

Prior to Gas Chromatography (GC) analysis, extracted cell wall lipids samples were subjected to metholysis [18] and analysed for FAMEs. Lipids samples were dissolved in 0.20 mL of Toulene and vortexed for few seconds. Methanol (1.5 mL) and 8% HCL solutions (0.3 mL) were added to the lipids solution. The mixtures were vortexed and incubated at 45˚C for 18 hours for mild methanolysis. To extract FAMES, the samples were cooled to room temperature and one mL of each hexane and miliQ water was added respectively to the samples. The mixtures were vortexed and hexane layers were directly analyzed by using GC after filtered through 0.22 µm PTFE filters.

2.5. Gas chromatography-mass spectrometry (GC-MS) analysis of FAMEs

FAMEs were analyzed by using GC-MS (Agilent, 5975C) with J & W GC column (30 m x 250 µm×0.25 µm; Agilent). The injector temperature maintained at 250°C. The column temperature was initially kept at 150°C for two mins and then increased at a rate of 12°C/min up to 200°C and held at this temperature for 15 mins. The temperature was increased to final temperature of 252°C and held for 10 mins at a rate of 3°C/min. Helium was used as the carrier gas at a flow rate of one mL/min and one µL of sample was injected using the splitless mode. The spectra were collected at 50-650 *m/z* range and were analyzed by using MZmine 2.21 software for mass detection, chromatogram builder and deconvolution. The detected peaks were subjected to compound identification using multiple online lipid and metabolomics database (NIST MS Search 2.0, METLIN, LipidMaps, KEGG and ChemSpider Databases).

3. Results and discussion

The findings from this study showed lipids composition of oil palm roots consists of FAMEs, sterols, neutral glycolipids and phospholipids. Full-scan spectra showed peak profiles were almost similar between control and *Ganoderma*-Infected roots. Approximately 56 compounds were detected in both control and *Ganoderma*-Infected roots as shown in table 1. These compounds are associated with normal metabolic activity of oil palm. Pentadecanoic acid, 14-methyl-, methyl ester (*m/z* 270.400; $C_{17}H_{34}O_2$) was detected at RT 16.66, which was the most abundant FAME in both control and *Ganoderma*-Infected cell wall roots. On the other hand, several compounds were detected at high abundance in *Ganoderma*-infected roots according to their base peak intensity data in Total Ion Chromatogram (TIC). Two fatty acid compounds; 3-methyl-6-(1-methyl-ethyl)-3, 9-decadien-1-ol (*m/z* 211.2; $C_{14}H_{26}O$) and 10, 13-Octadecadienoic acid, methyl ester (*m/z* 294.4; $C_{19}H_{34}O_2$) were more abundant in *Ganoderma*-Infected roots compared to controls. These compounds play significant roles in pathogen defence in response to various stresses [19]. A lipid phosphate derivatives bind with cyclic Guanosine Monophosphate (GMP), N'-Phosphoguanidinoethyl methyl phosphate (*m/z* 277; $C_4H_{13}N_3O_7P$) increased after the infection of *Ganoderma.* This compound is critically implicated in responses to both abiotic and biotic stress [20] such as gating of channels [21, 22], plant hormone signal transduction [23, 24], nitric oxide (NO)-dependent signalling [25] and regulation of transcription. Hence, the increase of this compound is proportional to the stress level in oil palm root tissue infected with *G. boninense.* Increment of phenolic compound, trans-trismethoxy Resveratrol

IOP Conf. Series: Earth and Environmental Science **77** (2017) 012014 doi:10.1088/1755-1315/77/1/012014

(*m/z* 270; $C_{17}H_{18}O_3$) in *Ganoderma*-infected roots also supports the indication of these compounds involvement in oil palm resistance against *Ganoderma* [11, 26]. Although the increment of hexadehydrovitamin D3/25-hydroxy-16,17,23,23,24,24-hexadehydrocholecalciferol (*m/z* 395.3; $C_{27}H_{38}O_2$) is not fully understood, this vitamin precursor might involve in cell-cell signalling to initiate defense mechanism in plant against stress and acts as an antioxidant compound to protect the plant from reactive oxygen species (ROS). Several lipids compounds were also detected in low amounts in roots infected with *G. boninense* (table 1).

Table 1. List of compounds that present in both control (healthy) and *Ganoderma*-Infected cell wall lipids of oil palm roots.

No	Compounds	RT (Min)	M/Z (30ppm tolerant)	Molecular Formula
1	2-methoxy-Ethanamine	6.87	74.900	C_3H_9NO
2	2-methyl-1-Pentene	8.01	85.000	C_6H_{12}
3	1,3,5-Trioxane	8.25	90.900	$C_3H_6O_3$
4	2-Propenal	8.29	56.900	C_3H_4O
5	1,2-Butadiene	8.46	54.900	C_4H_6
6	2-Propenamide	9.32	70.900	C_3H_5NO
7	* 3-methyl-6-(1-methyl-ethyl)-3,9-decadien-1-ol	11.38	211.200	$C_{14}H_{26}O$
8	p-methoxystilbene	11.46	211.100	$C_{15}H_{14}O$
9	2-Hexene	12.07	85.000	C_6H_{12}
10	* Propanenitrile	12.74	55.900	C_3H_5N
11	2-methyl-1,3-Butadiene	13.01	68.900	C_5H_8
12	* * Pentadecanoic acid, methyl ester	13.62	256.200	$C_{16}H_{32}O_2$
13	* N'-Phosphoguanidinoethyl methyl phosphate	14.63	277.000	$C_4H_{13}N_3O_7P_2$
14	* trans-trismethoxy Resveratrol	15.40	270.100	$C_{17}H_{18}O_3$
15	9-Hexadecenoic acid, methyl ester, (Z)-	16.29	268.100	$C_{17}H_{32}O_2$
16	* * Pentadecanoic acid, 14-methyl-, methyl ester	16.66	270.400	$C_{17}H_{34}O_2$
17	Benzenepropanoic acid, 3,5-bis(1,1-dimethylethyl)-4-hydroxy-, methyl ester	17.56	292.400	$C_{18}H_{28}O_3$
18	1-Butanol, 3-methyl-, acetate	18.06	114.900	$C_7H_{14}O_2$
19	1,2-Benzenedicarboxylic acid, diheptyl ester	18.30	148.800	$C_{22}H_{34}O_4$
20	5-Hydroxy-7-methoxy-8-C-methylflavanone	19.05	284.100	$C_{17}H1_6O_4$
21	* * Heptadecanoic acid, methyl ester	20.75	284.200	$C_{18}H_{36}O_2$
22	4-Methoxy potassium salicylate	23.69	207.000	$C_8H_8KO_4$
23	* 10,13-Octadecadienoic acid, methyl ester	24.38	294.400	$C_{19}H_{34}O_2$
24	* * 9-Octadecenoic acid, methyl ester, (E)-	24.66	296.400	$C_{19}H_{36}O_2$
25	* * 1-Dodecene	24.89	168.300	$C_{12}H_{24}$
26	* * Methyl stearate	25.80	298.500	$C_{19}H_{38}O_2$
27	3-Methoxymandelic acid-4-O-sulfate	26.67	278.000	$C_9H_{10}O_8S$
28	Benzenamine, N-phenyl-4-(phenylazo)-	27.75	273.300	$C_{18}H_{15}N3$
29	2(3H)-Furanone, dihydro-5-methyl-	28.20	100.100	$C_5H_8O_2$
30	(6Z,9Z,11E,13S)-13-Hydroperoxy-6,9,11-octadecatrienoic acid	28.99	310.200	$C_{18}H_{30}O_4$
31	* 2-Acetylthiazole	30.05	127.000	C_5H_5NOS
32	* *Octadecanoic acid, 2-hydroxy-, methyl ester	30.14	359.600	$C_{22}H_{47}O_3$

33	5-Hydroxy-7-methoxyisoflavone	30.17	269.100	$C_{16}H_{12}O_4$
34	Cyclopentanol	30.51	86.100	$C_5H_{10}O$
35	Heptane, 2,2,4,6,6-pentamethyl-	30.51	170.300	$C_{12}H_{26}$
36	Benzene, (1-methoxyethyl)-	31.25	136.000	$C_9H_{12}O$
37	2-chloromuconic acid	31.45	177.000	$C_6H_5O_4Cl$
38	Methyl 9.cis.,11.trans.t,13.trans.-octadecatrienoate	32.34	292.400	$C_{19}H_{32}O_2$
39	* * 2-oxo-4-methylthio-butanoic acid	32.47	149.000	$C_5H_8O_3S$
40	* * 2H-Pyran, 2,5-diethenyltetrahydro-	33.08	138.000	$C_9H_{14}O$
41	Eicosanoic acid, methyl ester	34.02	326.500	$C_{21}H_{42}O_2$
42	* * 2-Butenoic acid, methyl ester, (E)-	36.45	100.000	$C_5H_8O_2$
43	3-bromo-2Z-heptenoic acid	37.42	207.000	$C_7H_{11}BrO_2$
44	4-Acetoxy-3-methoxyacetophenone	37.90	160.000	$C_{11}H_{12}O_4$
45	* * 3,4-Hexanediol, 3,4-bis(4-hydroxyphenyl)-	37.97	302.000	$C_{18}H_{22}O_4$
46	gamma.-Sitosterol	39.52	414.700	$C_{29}H_{50}O$
47	* * 24-keto-25dehydrocholesterol	39.77	399.300	$C_{27}H_{42}O_2$
48	* * Docosanoic acid, methyl ester	40.56	354.600	$C_{23}H_{46}O_2$
49	stigmast-22E-en-3beta-o	41.22	414.400	$C_{29}H_{50}O$
50	* * 4-Hydroxy-2,3,3',4',5-pentachlorobiphenyl	42.39	339.900	$C_{12}H_5C_{15}O$
51	Tricosanoic acid, methyl ester	44.07	368.600	$C_{24}H_{48}O_2$
52	* 25-hydroxy-16,17,23,23,24,24-hexadehydrovitamin D3 / 25-hydroxy-16,17,23,23,24,24-hexadehydrocholecalciferol	49.48	395.300	$C_{27}H_{38}O_2$
53	2-Propenoic acid, ethenyl ester	50.70	100.100	$C_5H_8O_2$
54	Pentacosanoic acid, methyl ester	53.62	396.700	$C_{26}H_{52}O_2$
55	2-methylene-4-oxo-pentanedioic acid	54.04	159.000	$C_6H_6O_5$
56	Methyl 2-hydroxy-tetracosanoate	55.59	398.700	$C_{25}H_{50}O_3$

Note: * More abundant in infected tissues ** Less abundant in infected tissues based on base peak intensity in Total Ion Chromatogram (TIC).

In the present work, TIC chromatogram of control and *Ganoderma*-infected cell wall lipids revealed different peaks profiles as shown in figure 1.

Based on the GC-MS analysis, several peaks were not detected in *Ganoderma*-Infected roots and were only present in control oil palm cell wall lipids. The identified peaks are shown in table 2. After the roots were infected with *G. boninense,* these compounds may possible being suppressed or degraded root cells may unable to metabolize the respective compounds. Several factors might contribute to the suppression of metabolic activity in plant cell, which include stress factors and infection.

Table 2. List of compounds that were detected only in control (healthy) oil palm roots.

No	Compounds	RT (Min)	M/Z (30ppm tolerant)	Molecular Formula
1	Benzoic acid, 3-methyl-, methyl ester	15.96	152.000	$C_9H_{10}O_2$
2	Malonic acid	30.74	105.000	$C_3H_4O_4$
3	Methyl 2-hydroxy-eicosanoate	38.28	342.500	$C_{21}H_{42}O_3$
4	Tetracosanoic acid, methyl ester	48.34	382.700	$C_{25}H_{50}O_2$

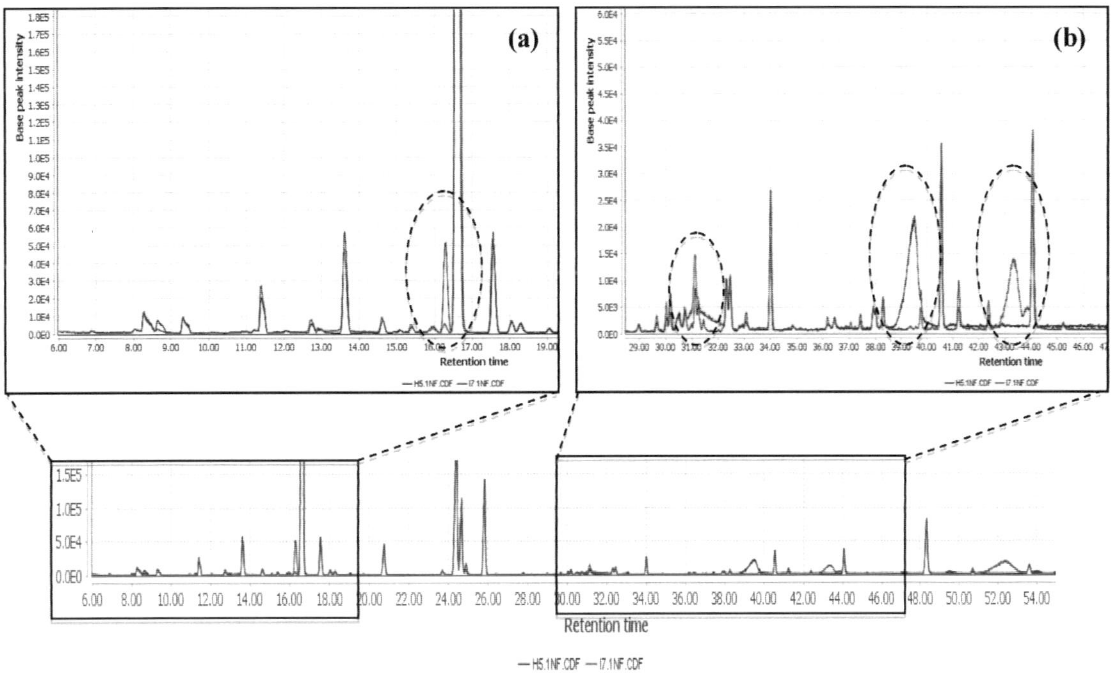

Figure 1. Overlayed GC-MS chromatogram of control (healthy) (blue line) and *Ganoderma*-infected (red line) oil palm roots cell wall lipids. (A) Enlarged area from RT 0-19 min. (B) Enlarged area from RT 29-47 min. Major different peaks are circled.

Based on the GC-MS analysis, several new peaks were detected and identified in the cell walls of *Ganoderma*-infected roots as summarized in table 3. However, these peaks were absent in healthy roots. These lipids may act as a key for plant defense response via signaling pathways of root tissues or play a key role in physiological changes and/or metabolic switch. Cyclohexane, 1, 2-dimethyl- is an aliphatic lipid component and by-product of lipid metabolism in plant. Increase of aliphatic compounds was reported previously in potato tuber infected by *Erwinia carotivora* [27].

Fatty acid methyl ester (FAME) compounds exhibited antimicrobial activity against pathogens. In this study, three FAMEs were identified as Methyl 2-hydroxy-16-methyl-heptadecanoate, 2-Propenoic acid, methyl ester and Methyl 9-oxohexacosanoate which possibly produced by oil palm roots to suppress *G. boninense* colonization. Similarly, previous study [28] found other fatty acids in oil palm tissue infected by *Ganoderma.* A heterocyclic aromatic compound, 5-[(3,7,11,15 Tetramethylhexadecyl)oxy] thiophene-2-carboxylic detected in oil palm tissues are usually produced by anaerobic microorganisms which exists endophytically in the plant tissues [29]. The presence of this compound in *Ganoderma*-infected tissue might indicate *G. boninense* as a facultative anaerobic fungi and lives endophytically in oil palm. Ergosta-5,7,22,24(28)-tetraen-3beta-ol is a derivative of fungal sterol found in the cell membrane. Thus, the presence of this compound confirmed *G. boninense* infection in root tissues. Previous studies suggested the importance of ergosterol in oil palm tissue for quantification of *Ganoderma* colonization [15, 30]. Another new compound detected after the infection of *Ganoderma*, is 7-Hydroxy-3',4'-methylenedioxyflavan, a benzopyran derivative and previously had been isolated from *Zephyranthes candida* [31]. It had been demonstrated that this compound in *Z. candida* was merely for defence and tolerant mechanism against biotic and abiotic stress. Therefore, this compound might play the similar role in *Ganoderma*-Infected root tissues. The presence of glycine in *Ganoderma*-Infected roots sample is quite fascinating. Previously, glycine-containing lipoamine acid was reported in *Cytophaga johnsonae* and its compound a combination of iso-3-hydroxy heptadecanoic acid, amide linked to glycine [32]. However, in this work glycine might be dissociated from its original gylcine-containing lipoamine acid structure due to the degradation by high temperature during GC analysis. This study is the first to report the presence of glycine in cell wall extract of oil palm roots infected with *G. boninense.* Glycine-containing lipoamine acid could also be developed as biomarker for BSR early detection. A flavonoid compound, (S)-4'-Hydroxy-4-methoxydalbergione, which is lipid-soluble, was detected in cell wall lipids extract of *Ganoderma*-Infected roots. This compound plays important role in plant resistance against pathogens. Flavonoid compounds are transported to the site of infection and induce the hypersensitivity reaction, which is the earliest defense mechanism employed by the infected plants and programmed cell death [33]. This compound might accumulate in cell walls during interaction between oil palm and *G. boninense.* Previous study also found the accumulation of flavonoid compounds in cotton cell wall after inoculated with *Xanthomonas campestris* pv. *malvacearum* [34].

Table 3. List of compounds which were detected only in *Ganoderma*-Infected oil palm roots samples.

No	Compounds	RT (Min)	M/Z (30ppm tolerant)	MF
1	Cyclohexane, 1,2-dimethyl-	29.66	112.000	C_8H_{16}
2	Methyl 2-hydroxy-16-methyl-heptadecanoate	31.11	299.500	$C_{18}H_{35}O_3$ -
3	2-Propenoic acid, methyl ester	36.18	86.000	$C_4H_6O_2$
4	Methyl 9-oxohexacosanoate	37.05	424.400	$C_{27}H_{52}O_3$
5	5-[(3,7,11,15-Tetramethylhexadecyl)oxy]thiophene-2-carboxylic acid	37.23	424.300	$C_{25}H_{44}O_3S$
6	Ergosta-5,7,22,24(28)-tetraen-3beta-ol	43.34	394.200	$C_{21}H_{28}O_3$
7	7-Hydroxy-3',4'-methylenedioxy flavan	47.11	271.100	$C_{16}H_{14}O_4$
8	Glycine	52.39	75.000	$C_2H_5NO_2$
9	(S)-4'-Hydroxy-4-methoxydalbergione	55.1	271.100	$C_{16}H_{14}O_4$

4. Conclusion

In brief, this work presented a comparison of cell wall lipids composition between the healthy and *Ganoderma*-infected oil palm root tissue extracts. Through GC-MS analysis, some lipids compounds were only detected in *G.boninense* infected oil palm roots. New lipids and some lipid-linked compounds associated with plant defense response were discovered in this study. The fluctuation of several compounds between healthy and *Ganoderma*-Infected root tissue were also revealed. Further investigation is needed to explore the roles or mechanism of action of newly found compounds during pathogenesis of *Ganoderma* in oil palm. These compounds has valuable potential for development of biomarker and that might lead to better control strategies for combating *Ganoderma* infection in oil palm.

References

[1] Guni O B, Naga J A and Chong K P 2012 The potential of chitosan in suppressing *Ganoderma boninense* infection in oil palm seedlings *J. Sus. Sci. Mgmt* **7(2)** 186-92
[2] Arif M A M, Arbizah B Y, Zetty Norhana S, Syahanim, Idris A S, Mohd Din A and Ravigadevi S 2007 Molecular and biochemical approaches to understanding oil palm-*Ganoderma* Interactions *Proc. Agri. Biotech. Sustain. Conf.* (Malaysia: Malaysia Palm Oil Board-MPOB)
[3] Corley R H V and Tinker P B 2003 *The Oil Palm* (USA:Blackwell Publishing Company)
[4] Carpita N C and McCann M 2000 The Cell Wall *Biochemistry and Molecular Biology of Plants* eds B Buchanan, W Gruissem and R L Jones (Rockville, MD: American Society of Plant Physiologists) pp 52-108
[5] Almagro L, Gómez Ros L V, Belchi-Navarro S, Bru R, Ros Barceló A and Pedreño M A 2009 Class III peroxidases in plant defence reaction *J. Exp. Bot.* **60(2)** 377-90
[6] Chisholm S, Coaker G, Day B and Staskawicz B J 2006 Host-microbe interactions: shaping the

evolution of the plant immune response *Cell* **124** 803-14

[7] Collinge D B 2009 Cell wall appositions: The first line of defense *J. Exp. Bot.* **60** 351-2

[8] Schwessinger B, Ronald P C, Schwessinger B and Ronald P C 2012 Plant innate immunity: Perception of conserved microbial signatures *Annu. Rev. Plant Biol.* **63** 451-82

[9] Siebers M, Brands M, Wewer V, Duan Y, Hölzl G and Dörmann P 2016 Lipids in plant–microbe interactions *Biochim. Biophys. Acta.* **1861(9)** 1379-95

[10] Shah J and Chaturvedi R 2009 Lipids in plant–microbe interactions *Annl. Plant Rev.* **34** 292-333

[11] Mohamad Arif A M, Abrizah O, Norhana B Y Z, Syahanim S, Idris A S, Din A M and Ravigadevi S 2007 Molecular and biochemical approaches to understanding oil palm-*Ganoderma* interactions *Proc. Agri. Biotech. Sustain. Conf.* (Kuala Lumpur, Malaysia) pp 228-47

[12] Penson S P, Schuurink R C, Fath A, Gubler F, Jacobsen J V and Jones R L 1996 cGMP is required for gibberellic acid-induced gene expression in barley aleurone *Plant Cell* **8** 2325-33

[13] Chong K P, Lum M S, Foong C P, Wong C M V L, Atong M and Rossall S 2011 First identification of *Ganoderma boninense* isolated from Sabah based on PCR and sequence homology *Afr. J. Biotechnol.* **(10)** 14718-23

[14] Ariffin D and Idris A S 1992 The *Ganoderma* Selective Medium (GSM) *PORIM Information Series* vol 8 (Malaysia: Palm Oil Research Institute of Malaysia-PORIM)

[15] Chong K P 2012 An evaluation of the *Ganoderma* fungal colonisation using ergosterol analysis and quantification *The Planter* **88(1034)** 311-9

[16] Feiz L, Irshad M, Pont-Lexica R F, Canut H and Jamet E 2006 Evaluation of cell wall preparations for proteomics: A new procedure for purifying cell walls from Arabidopsis hypocotyls *Plant Methods* **2** 10

[17] Folch J, Lees M and Stanley G H S 1957 A simple method for the isolation and purification of total lipides from animal tissues *J. Biol. Chem.* **226** 497-509

[18] Ichihara K and Fukubayashi Y 2010 Preparation of fatty acid methyl esters for Gas-Liquid Chromatography *J. Lipid Res.* **51(3)** 635-40

[19] Varns J L and Glynn M T 1979 Detection of disease in stored potatoes by volatile monitoring *Am. Potato J.* **56** 185-97

[20] Gottig N, Garavaglia B S, Daurelio L D, Valentine A, Gehring C and Orellano E G 2008 *Xanthomonas axonopodis* pv. citri uses a plant natriuretic peptide-like protein to modify host homeostasis *Proc. Natl. Acad. Sci. USA* **105** 18631-36

[21] Hoshi T 1995 Regulation of voltage dependence of the KAT1 channel by intracellular factors *J. Gen. Physiol.* **105** 309-28

[22] Leng Q, Mercier RW, Yao W and Berkowitz G A 1999 Cloning and first functional characterization of a plant cyclic nucleotide-gated cation channel *Plant Physiol.* **121** 753-61

[23] Jocelyn F T 1905 The constitution of glutaconic acid *J. Chem. Soc. Trans.* **87** 1669-85

[24] Neill S J, Desikan R and Hancock J T 2003 Nitric oxide signalling in plants *New Phytol.* **159** 11-35

[25] Chong K P, Rossall A and Atong M 2011 HPLC fingerprints and *In vitro* antimicrobial activity of syringic acid, caffeic acid and 4-hydroxybenzoic acid against *Ganoderma boninense* *J. Appl. Sci.* **11** 2284-91

[26] Ghosal S, Singh S K and Srivastava R S 1985 Flavans from *Zephyranthes* flava *Phytochemistry* **24** 151-3

[27] Sapak Z, Meon S and Ahmad Z A M 2008 Effect of endophytic bacteria on growth and suppression of *Ganoderma* infection in oil palm *Int. J. Agri. Biol.* **10** 127-32

[28] Rozlianah F S, Jualang A G and Chong K P 2015 Fatty acids and phenols involved in resistance of oil palm to *Ganoderma boninense* *Adv. Environ. Biol.* **9(7)** 11-6

[29] Kawazoe R, Okuyama H, Reichardt W and Sasaki S 1991 Phispholipid and a novel-containing lipoamino acid in *Cytophaga johnsonae* stanier strain C21 *J. Bacteriol.* **173(17)** 5470-5

[30] Alexander A and Chong K P 2014 *Early Detection of Ganoderma in Oil Palm: A field Guide* (Malaysia: Universiti Malaysia Sabah)

[31] Pharmawati M, Maryani M M, Nikolakopoulos T, Gehring C A and Irving H R 2001 Cyclic GMP modulates stomatal opening induced by natriuretic peptides and immunoreactive analogues *Plant Physiol. Biochem.* **39** 385-94

[32] Mierziak J, Kostyn K and Kulma A 2014 Flavonoids as important molecules of plant interactions with the environment *Molecules* **19** 16240-65

[33] Dai G H, Nicole M, Andary C, Martinez C, Bresson E, Boher B, Daniel J F and Geiger J P 1996 Flavonoids accumulate in cell walls, middle lamellae and callose-richpapillae during an incompatible interaction between *Xanthomonas campestris* pv. *malvacearum* and cotton *Physiol. Mol. Plant Pathol.* **49** 285-306

[34] Tanaka H, Maeda H, Suzuki H, Kamibayashi A and Tonomura K 1982 Metabolism of Thiophene-2-carboxylate by a photosynthetic bacterium *Agri. Biol. Chem.* **46(6)** 1429-38

A study on the spatial characteristics and correlation of migrant workers' urban integration and well-being: A case study of Xi'an (China)

D H Wang[1], X J Yang[2, 4] and F J Hao[2, 3]

[1] School of Information Management and Statistics, Hubei University of Economics, Wuhan, 430205, China
[2] Tourism School-Human Geography Institute, Xi'an International Studies University, Xi'an, 710128, China
[3] Hubei Local Taxation Research Center, Wuhan, 430205, China

E-mail: wdhong@hbue.edu.cn

Abstract. This paper used SPSS and ARCGIS to measure the urban integration degree and well-being index, spatial features, and their correlation. This results show: (1) The space differentiation of migrant workers' urban integration degree in Xi'an distinct: The northern great site protection zone area is low, eastern military area is peak and the western electronic district and southwest high-tech zone are second peak areas. (2) Migrant workers' well-being index has differentiation spatial distribution: eastern military area is significantly higher than other regions, northern economic zone shows low-lying shape, southern cultural and educational area is higher than northern economic development zone, and central business district is higher than the surrounding. (3) As the result of correlation analysis in SPSS 19.0, it is shown that there is certain positive correlation between urban integration degree and well-being index of migrant workers in main urban districts of Xi'an. Economic integration and social integration have positive prediction to well-being.

1. Introduction

Migrant workers are special social product of urbanization in China. In 2016, China's urbanization rate has reached 57.35%, and the number of migrant workers reached 281.11 million, of which 169.34 million rural workers employed outside their hometowns. China's urbanization is inseparable from the great contribution of migrant workers [1]; they are divorced from the edge of rural and urban social system, which is passive urbanization, pseudo-urbanization, incomplete urbanization [2]. The function of the city gradually changed from "economic orientation" to "well-being orientation". Welfare, health care [3], efficiency and fairness become new problems of new economic geography [4, 5].

The integration measurement of urban migrant workers did not form a unified standard, which is generally believed that urban integration includes four dimensions: social, cultural, economic and political integration [6], involving economic, social and psychological subjective and objective multidimensional [7, 8]. Empirical studies have shown that degree of integration of migrant workers in northern China is higher than that in the southern cities [9] and as for the western region of migrant workers into the city, there is also spatial differentiation [10].

The theory of “Maslow's hierarchy of needs” argues that human needs include physical, security, social acceptance, esteem, and self-actualization: our desires to live up to our unique potentialities and our needs for self-fulfillment [11]. The well-being of migrant workers is based on the individual's own value standards to make a holistic assessment of the quality of life [12]. The Positive social integration brings strong subjective well-being to immigrants [13, 14]; social integration is conducive to the promotion of migrants' physical health and mental health [15]. Social support [16], satisfaction of relatives and friends [17-19], dwelling environments [20], the subjective well-being of immigrants have a significant positive correlation. Social capital and social participation exactly play a decisive role in the well-being of immigrants [21]. On the contrary, it will weaken the level of well-being [22]. Foreign scholars believe that social support in social integration has a positive effect on the well-being of immigrants, and will buffer the negative impact of adverse events. More extensive research in this area has certain reference to domestic research inspiration.

Social integration and economic integration have a significant impact on the willingness of migrant workers' development. Psychological integration and cultural integration have significant effects on the mental health of migrant workers [23]. Empirical studies show that high degree of integration of migrant workers feel that society is more equitable [24, 25]. Social capital and social relations clearly affect subjective well-being [26, 27]. The academic community has studied the sense of well-being from the factors influencing the urban integration factors such as working hours and social capital, but few studies have made a comprehensive analysis of the well-being of migrant workers in the dimensions of urban integration.

2. Data and methodology

The metropolis can provide more jobs, but migrant workers in the city's housing, schooling, health care and other costs are very high. Coupling with a short period of time cannot solve the city’s urban disease but hinder the citizenization of migrant workers [28]. The number of migrant workers in Xi'an is more than one million, and has been increasing year by year. Migrant workers into the city after the identity change, identity reconstruction, social network reconstruction, entertainment and other dilemmas with the corresponding spatial response to the performance of economic space, social space, cultural space, psychological space, recreational space reconstruction. This paper refers to migrant workers in the geographical space, from rural areas into the city area, in the household, the household registration of migrant workers without city, from the primary sector to secondary sector or tertiary sector space, and living in the city with low income, no benefits, and no guarantee of belonging to vulnerable groups in the city. The sense of well-being of migrant workers refers to the overall perception and evaluation of urban social life by the way of subjective and objective conditions in a certain social and economic environment.

According to the field survey of the main urban area of Xi'an, the specific investigation sites are small living village, cultural road, Yuhua village, Zhangjiapu, Taihua South Road, Dizhai and Tumen. The concentration of migrant workers in these areas is high, space coverage is wide and the type of research site is diversified. The research areas represent the migrant workers' suburbs, the self-managed labor market, the self-diversified labor market and the government's labor market. A total of 550 questionnaires were distributed and 475 were collected. Excluding incomplete questionnaires, we received a total of 446 valid questionnaires, and the effective utilization of the questionnaire was 89.2%. The proportion of agricultural household registration accounted for 99.55% of the effective questionnaire in line with the identity of the study. After statistical analysis, the reliability and validity of the samples reached the requirements.

Based on the purpose of the study and the relevant information collected, the urbanization and well-being of migrant workers were selected from five dimensions respectively and 15 individual indicators to construct the index system. SPSS software was used for statistical analysis to verify the existence of migrant workers' urbanization and well-being index correlation.

3. Analysis of the results

By principal component analysis, we selected the factor which eigenvalues greater than 1, according to principal component score coefficient, and the proportion of each of the principal components of the total variance, calculated degree of migrant workers' urban integration and well-being index comprehensive score, respectively. The spatial location data (geographical coordinates) of the individual migrant workers are acquired by the survey which is associated with the comprehensive score. We used ArcGIS10.1 spatial statistical analysis to map the degree of integration of migrant workers' cities and the attributes of well-being index.

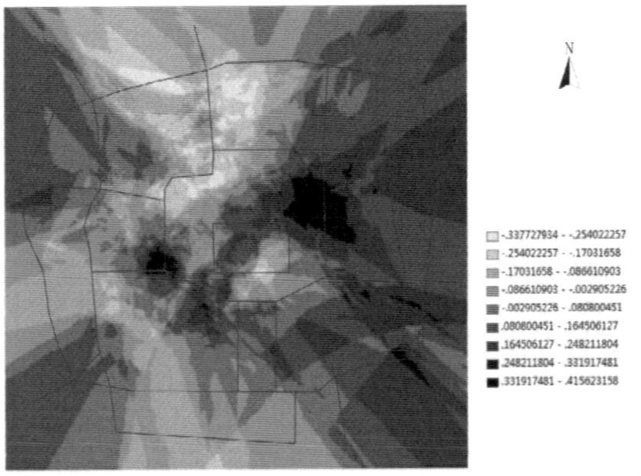

Figure 1. Spatial pattern of migrant workers' urban integration degree.

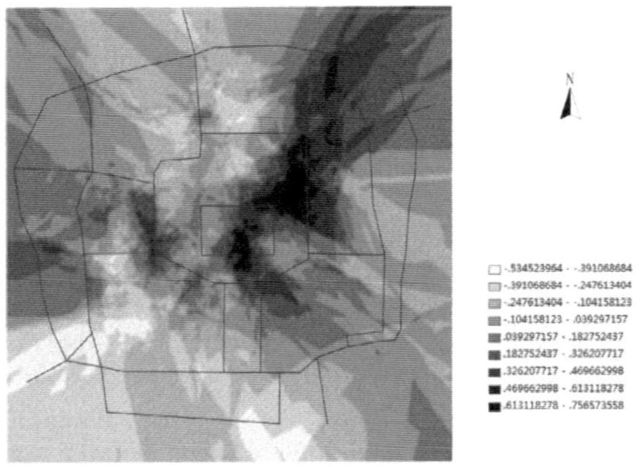

Figure 2. Spatial pattern of migrant workers' well-being index.

Figure 1 means that the degree of spatial integration of migrant workers in Xi'an is obviously different: The eastern part of the military sites belongs to the low value area, the eastern military city belongs to the high value area and the western suburbs electronic city, the southwest high-tech zone belongs to the secondary high value area. Figure 2 shows that the spatial distribution of migrant workers' well-being index in Xi'an is non-uniform; the well-being index of migrant workers in the eastern military city of Xi'an is obviously higher than that in other regions, and the well-being index of migrant workers in the northern economic development zone is low. The well-being index of migrant workers in the southern cultural and educational district is higher than that in the northern economic development zone, and the well-being index of the central business district is higher than that of the surrounding area.

SPSS 19.0 software was used to analyze the correlation between the total value of the migrant workers' urban and the total value of well-being by the Pearson correlation coefficient. The concrete results are shown in table 1.

Table 1. The correlation of urban integration and well-being.

		Urban integration score	Well-being score
Urban integration score	Pearson Correlation	1	.311**
	Sig.(2-tailed)		**.000**
	N	446	446
Well-being score	Pearson Correlation	**.311****	1
	Sig.(2-tailed)	.000	
	N	446	446

The significant score of the total score of urban integration and the gross score of well-being is 0.000 and the correlation coefficient is 0.311, which indicates that there is a positive correlation between urban integration and well-being, which proves that the hypothesis of the article has positive correlation.

Table 2. The correlation between five dimension and well-being.

		Economic integration	Social integration	Political integration	Cultural integration	Psychological integration
well-being	Pearson Correlation	.423*	.167*	.138**	.004	.035
	Sig.(2-tailed)	**.000**	**.000**	.003	.933	.456
	N	446	446	446	446	446

As shown in table 2, the significant value between the dimensions of economy and society with well-being is 0.000, which is less than the significance level of 0.01, demonstrating that the integration dimensions of economy and society is correlated with well-being. Correlation coefficients are 0.423, 0.167, respectively, which indicate that dimensions of economy, society integration have positively relationship with well-being. Among them, the economy integration dimension has the highest correlation following by the society integration. Dimensions of politics, culture, and psychology integration have not significant linear correlation relationship with well-being.

The sense of well-being of migrant workers is the microcosmic individual comprehensive perception of diverse demand satisfaction, the degree of factors affecting importance is different. The economy dimension has most influence on migrant workers' well-being, which is due to the economic demands, and is the primary reason for migrant workers getting into urban. Migrant workers work in cities all the year round, so most of them take city dwellers as reference. Only when migrant workers have perceived the improvement of employment, housing, income level, which can improve their quality of life in city, can they enjoy the same results of urbanization with urban residents, thereby promoting their well-being.

Social integration dimension takes second place in influencing migrant workers' well-being. The original acquaintance social network is gradually being broken after the migrant workers enter into the city, only after the system of social network is rebuilt, can they get social support. The more contacts with local residents, the more social support they felt. On the contrary, the greater gap between social distance, the stronger sense of being marginalized and discriminated, weakens migrant workers' satisfaction of the quality of urban social space, hindering their urban integration, which will affect their perception of well-being.

As for political integration dimension, migrant workers are lack of consciousness of political participation, lack of enough attention for social security, and lack of knowledge of the rights and interests protection ways. The human capital of migrant workers is low, the instability of work, the lack of legal knowledge, the deprivation of the employment unit, make them lack of contract and insurance protection. Migrant workers have not discovered their political rights and social security consciousness, so political integration effect on well-being does not have explicit linear correlation.

Psychological integration is the deep dimension of urban integration, and it has not significant linear impact on well-being. Migrant workers enter into city to seek long-term development, facing all sorts of difficulties, but they have certain psychological preparation. They are lack of sense of security, identity, and they are not sensitive to psychological distance and discrimination awareness, leading to their psychological perception level has no obvious impact on well-being.

Cultural integration dimension reflects the sense of comfort and identity of the city for the migrant workers. Cities are bound to provide more life convenience and high level of service quality than rural. Generally, migrant workers are yearning for city life, and they have the same urban cultural tradition with urban dwellers, so cultural integration does not have significant linear effect on well-being.

Use the R value to test the degree of independence of the independent variable and the dependent variable, and the range of values is between 0-1, the closer to 1, the more closely linked. Table 3 shows that the fitting model of urban integration and well-being which is preferable.

Table 3. Model Summary.

Model	R	R Square	Adjusted R Square	Sed. Error of the Estimate
1	.472[a]	.223	.214	.526622754360832

[a]Predictors: (constant), Economic integration, Social integration, Cultural integration, Political integration, Psychological integration.

By means of variance analysis to verify whether the regression equation is established. If the probability of significance (Sig) <0.05, the equation is established, otherwise invalid. Table 4 shows that the equation is established.

Table 4. Anova[b].

		Sum of Squares	DF	Mean Square	F	Sig.
1	Between groups	34.932	5	6.986	25.192	.000a
	Within groups	122.026	440	.277		
	total	156.958	445			

[b] Dependent Variable: Well-being score

Table 5. Coefficient[a].

Model	Unstandardized Coefficient		Standardized Coefficient	t	Sig.
	B	Std. Error	Beta		
1(Constant)	2.088E-7	.025		.000	1.000
Economic integration	.252	.025	.425	9.952	.000
Social integration	.105	.025	.177	4.165	.000
Political integration	.046	.025	.078	1.828	.068
Cultural integration	.006	.025	.011	.250	.803
Psychological integration	.041	.025	.069	1.632	.103

[a] Dependent Variable: Well-being score

Table 5 shows the coefficient of the regression model, in which the value of the standard coefficient indicates the influence of the each independent variable on the dependent variable. The Sig values of economic integration and social inclusion are both 0.000 and below 0.05, indicating a predictive effect on the well-being of migrant workers. The regression coefficients are 0.252 and 0.105 respectively, indicating that economic integration and social integration are well-being Sense has a positive predictive effect. And the coefficient Sig of the constant term is 1> 0.05, which is not verified by significance test. The standard regression equation of migrant workers' well-being is:

Migrant workers well-being = 0.425 *economic integration + 0.177 * social integration + error coefficient

The results show that the economic integration and social integration of migrant workers in Xi'an are the main factors influencing the well-being, and the urban integration has positive predictive effect on well-being.

4. Conclusion

The degree of spatial integration of migrant workers in Xi'an is obviously different: The eastern part of the military sites belongs to the low value area; the eastern military city belongs to the high value area; the western suburbs electronic city, the southwest high-tech zone belongs to the secondary high value area;

The spatial distribution of migrant workers' well-being index in Xi'an is non-uniform; the well-being index of migrant workers in the eastern military city of Xi'an is obviously higher than that in other regions, and the well-being index of migrant workers in the northern economic development zone is low. The well-being index of migrant workers in the southern cultural and educational district is higher than that in the northern economic development zone, and the well-being index of the central business district is higher than that of the surrounding area.

The economic integration and social integration of migrant workers in Xi'an are the main factors influencing the well-being, and the urban integration has positive predictive effect on well-being.

Acknowledgments

This research received grants from the National Social Science Foundation of China (13XSH017).

References

[1] Cai F 2016 *Chinese Rural Disc.* **34(1)** 14-9
[2] Cai F 2014 *Urban Environ. Stud.* **2(2)** 3-12
[3] Purohit B C 2016 *Rev. Urban Reg. Dev. Stud.* **28(1)** 55-74
[4] Hochstenbach C and Boterman W R 2017 *Urban Stud.* **54(2)** 399-420
[5] An H S 2009 *Principles of New Economic Geography* (Beijing: Economic Science)
[6] Entzinger H and Biezeveld R 2003 *Benchmarking in Immigrant Integration* (Rotterdam: Erasmus University Rotterdam)
[7] Yang J H 2010 *Popul. Econ.* **179(2)** 64-70
[8] Zhou H 2012 *Popul. Res.* **36(3)** 27-37
[9] Tian M and Bo J L 2014 *Hum. Geogr.* **29(1)** 43-8
[10] Wang D H, Li K Y and Hao F J 2014 *Northwest Popul. J.* **35(6)** 64-70
[11] Coyle-Shapiro J and A-MConway N 2005 *J. Appl. Psychol.* **90(2)** 774-81
[12] Diner E 1984 *Psychol. Bull.* **95 (03)** 542-75
[13] Ward C 2013 *Int. J. Intercult. Rel.* **37(04)** 391-404
[14] Angelini V, Casi L and Corazzini L 2015 *J. Popul. Econ.* **28(03)** 817-44
[15] Sam D L and Berry J W 2010 *Perspect Psychol. Sci.* **5(04)** 472-81
[16] Sarracino F 2010 *J. Socio-Econ.* **39(4)** 482-517
[17] Chou K L 1999 *J. Genet. Psychol.* **160(3)** 319-31
[18] Feng J and Prillwitz J 2013 *Urban Studies* **50(14)** 2993-3010
[19] Zhao Y, Dijst M and Chai Y W 2016 *Urban Stud.* **53(12)** 2469-87
[20] Ishikawa, Noriko and Fukushige 2015 *Rev. Urban Reg. Dev. Stud.* **27(03)** 149-76
[21] Cooper H, Okamura L and Gurka V 1992 *Pers. Indiv. Differ.* **13(5)** 573-83
[22] Mui A C 2001 *J. Hum. Behav. Soc. Environ.* **3(4)** 281-99
[23] Yue Z S, Li S M and Jin X Y 2011 *Chinese J. Socio.* **31(6)** 130-52
[24] Nie W and Feng X T 2013 *J. Nanjing Agr. U.* **53(05)** 32-40
[25] Hu R and Cheng S S 2010 *J. Xiamen U.* **200(4)** 97-105
[26] Chang W 2009 *Int. J. Soc. Econ.* **36(8)** 844-68
[27] Xing Z J and Zhang Y 2007 *Soc. Sci. Res.* **30(6)** 1436-8
[28] Fang C L 2014 *Geogr. Res.* **33(4)** 674-86

Fluctuation traits of Litchi wholesale price in China

F F Yan[1], W E Qi[1,2,3] and X Ouyang[1]

[1] Department of Economics and Management, South China Agricultural University, Guangzhou, Guangdong, 510642, China
[2] National Litchi & Longan Industry's Technology System, Guangzhou, Guangdong, 510642, China

E-mail: qiwene@scau.edu.cn

Abstract. This paper chose the wholesale price of litchi as research object based on the daily data of 11 main sales markets in China -- Beijing, Chengdu, Guangzhou, Hefei, Jiaxing, Nanjing, Shanghai, Shenyang, Changsha, Zhengzhou and Chongqing from April 1, 2012 to September 30, 2016. After analyzing the fluctuation characteristics with BP filter method and H-P filter method, and the fluctuation trends of litchi wholesale price in China obtained by BP filter are roughly consistent with the trends obtained by H-P filter. The main conclusions are as follows: there is strong cyclicality in the fluctuation of litchi wholesale price; the period of fluctuations of litchi wholesale prices are not repeatable; litchi wholesale price fluctuates asymmetrically in one fluctuation cycle.

1. Introduction

In 2016, there were about 1,853,377 acres of planting area of litchi all over the world. And there were 1,546,952 acres in China, accounting for 80 % of the world's litchi planting area and ranked 1. In 2016, there were about 2,012,900 tons of litchi all over the world. And the total output of litchi in China is 1,702,100 tons, accounting for 84.56% of the whole world. Therefore, it is of great importance to study the Chinese litchi industry as it plays a decisive role in the world economy. According to the Statistics Database of PRC, planting area of litchi ranks 5, just following citrus, apples, pears and grapes, the top four economic agro-products in China. Therefore, it is of practical significance to study litchi industry.

Besides, in term of the property, litchi is a kind of typically perishable fruit, so it has a short shelf-period; and litchi is vulnerable to some factors such as season, climate conditions, supply and demand, so there tends to be great risk of litchi market and big fluctuations of litchi price, affecting the profits of litchi producers, consumers and other stakeholders in the supply chain. Therefore, it is of great significance to analyze the fluctuation characteristics of litchi price, because it can guide farmers and relative producers to arrange the production, grasp the rhythm of the operation and increase their income -- beef up the benefits of consumers.

There are four main planting provinces in China, namely Hainan, Guangdong, Guangxi Zhuang Autonomous Region and Fujian. Litchi is one of the main cash crops in these provinces. Litchi industry provides employment and survival opportunities for stakeholders in Hainan, Guangzhou, Guangxi and Fujian. According to incomplete statistics, there are 3,288 households engaged in litchi cultivation in the four provinces, and each household possesses approximately 9.23 acres of litchi garden for average. Litchi production is mainly for sale, so farmers rarely eat their own. According to rough calculation, in 2016, the total turnover in litchi wholesale market was up to 300 billion RMB. That is why litchi industry makes a great contribution to Chinese Gross Domestic Product. Litchi sales in China mainly concentrate in six areas in China -- the Northeast market, the North market, the Central market, the South market, the East market and the Southwest market, covering more than 90% of the litchi sales markets in China. Therefore, when studying the price fluctuation of litchi in China, it is representative to choose litchi whole price of the Northeast market, the North market, the Central market, the South market, the East market and the southwest market as research object.

In recent years, a large number of literatures have studied the characteristics of price fluctuation. The main conclusions are as follows: the prices of agro-products fluctuate differently with different varieties [1]; in spatial transmission ,the price of agro-products has significant heterogeneity [2]; and agricultural price transmission is asymmetric, namely the price swings like "rocket and feather" [3-22]. The theoretical contribution of this paper is that it expanded the economic research of litchi industry in China. As litchi is different from food crops and some vegetables or other bulk fruit like bananas, apples, grapes and pears which are necessity goods in our daily life, litchi has not been placed in an important position all the time, which leads to an absence of litchi price data on the Statistics Database of PRC and makes the researches on litchi industry be not profound enough in academic community. But as is explained above, it is certainly worth focusing more on the litchi industry, especially the fluctuation of litchi price. So, this paper tries to analyze the price fluctuation by BP filter and H-P filter so as to enrich the research content of litchi economy and verify the theory of asymmetric transmission, because the analysis is based on the theory of asymmetric transmission.

The practical contribution of this paper is to provide corresponding measures to avoid management risk in litchi industry, to resist the impact of price fluctuation and to ensure that the litchi industry can get a healthy and sustainable development by analyzing the characteristics of price fluctuation in litchi market.

From the perspective of data, most data types on agro-product price fluctuation are quarterly data, monthly data or weekly data, while it is rare to use daily data to analyze the price volatility. This paper uses daily data to study the fluctuation characteristics of wholesale price in the litchi markets, discloses the traits hidden in the quarterly data, monthly data or weekly data by average the daily volatility, which might be another contribution of this paper. In addition, this paper uses both BP filter and H-P filter to analyze the price fluctuation. Because there is no requirement for the length of the economic time series using H-P filter. However, on the one hand, it is relatively free to choose the parameters for H-P filter method, resulting in uncertainty of the economic fluctuations after H-P filter. On the other hand, the unrelated economic sequence may have a significant correlation after H-P filter [23], and if the economic time series exhibits smoothness after first order difference, may induce pseudo-periodic phenomena [24]. While BP filter method has a high requirement for the length of the economic time series, that is, there is limitation for small sample sequence, but it can overcome the problem of parameter selection problem of H-P filter, because BP filter method requires the level stable or first order stable of the economic sequence. Therefore, in order to make more accurate analysis of litchi wholesale price fluctuation, this paper uses BP filter and H-P filter and compares the results of the two filters and speculates that litchi wholesale price may be weakly dependent on filer models.

The design of this paper is as follows:

- First, use BP filter method to analyze litchi wholesale price of 11 sales markets;
- Second, analyse the samples by H-P filter;
- Third, compare the analysis of BP filter and H-P filter on price fluctuations of litchi wholesale;
- At the end, analyze the possible reasons of price fluctuations.

2. Research methods

2.1. Variable selection and sample source

Similar to most of the fresh agro-products, although there are new sales channels such as e-commerce, the main channels of litchi sales are still the traditional wholesale and retail. Usually, litchi distributors procurement litchi goods through the wholesale sites, the wholesale markets or contracting orchard directly in the origins, and then transport them to the wholesale markets for sale, and then retail them to terminal consumers via various retail methods. Therefore, the wholesale price is the most important indicators in the circulation of litchi, and is a barometer in litchi market. Changes in the wholesale price will directly lead to Domino effect of the wholesale price of the upstream origins and the retail price of the downstream. Therefore, this paper chooses litchi wholesale price as the main variable.

The data of this paper are derived from "Basic Database of National Litchi & Longan Industry's Technology System", which covers data in 11 major sales market of litchi including Northeast China market (Shenyang), North China market (Beijing), Central China market (Changsha, Zhengzhou), South China market (Guangzhou), East China market (Shanghai, Jiaxing, Nanjing and Hefei) and Southwest market (Chongqing, Chengdu). In this paper, we selected the daily data of litchi wholesale price from April 1, 2012 to September 30, 2016. The total number of samples is 5195.

2.2. Model establishment
On the study of price fluctuation of agro-products, previous literature usually used the original time series, but few used the trend data after decomposing original series into trend parts, seasonal parts and irregular parts. The common method to decompose data is seasonal adjustment method, which is only suitable for monthly data or quarterly data but not daily data. So, this article does not use the season adjustment method, but use the filter method directly to study the litchi wholesale price fluctuations. The main filter methods of decomposing price of agro-products are H-P filter [25-27] and BP filter. In order to get more accurate results, this paper uses both BP filter method and H-P filter method to analyze litchi wholesale price and compare the results of these two methods.

Baxer and King[28] proposed BP filter method based on spectral analysis of the decoupling method, also known as spectral analysis, the modelling idea: time sequence $X = \{x_1, x_2, ..., x_t\}$, T is the length of time series X. The main idea of spectrum analysis sees the fluctuation of the economic time series X as the results of several different frequencies, and tries to find out the most explanatory of the fluctuation of the economic event sequence X in these different frequencies.

Hodrick and Prescott [29] proposed an H-P filter method to study the fluctuation of the economic sequence. The idea of this method is similar to that of BP filter, except that the H-P filter method considers the fluctuation of the economic time series as the superposition of the growth component and the periodic component. The use of the method has a long history, especially in actuarial science. The conceptual framework of the H-P filter method is as follows: A given economic time series $\{Y_t\}$ is the summation of the growth component $\{Y_t^T\}$ and the periodic component $\{Y_t^C\}$, then the specific expression is:

$$Y_t = Y_t^T + Y_t^C \tag{1}$$

The smoothing method of the H-P filter method for the economic time series $\{Y_t\}$ fluctuation is the square sum of the second order difference of the economic event sequence. Then the periodic component $\{Y_t^C\}$ of the economic sequence $\{Y_t\}$ is separated from the growth component $\{Y_t^T\}$. Since the conceptual framework of the method is long-term, the average of the long-term periodic component $\{Y_t^C\}$ tends to zero. Based on above conditions, growth ingredients are as follows:

$$\min_{\{Y_t^T\}_{t=-1}^{T}} \left\{ \sum_{t=1}^{T} Y_t^{C2} + \lambda \sum_{t=1}^{T} [(Y_t^T - Y_{t-1}^T) - (Y_{t-1}^T - Y_{t-2}^T)]^2 \right\} \tag{2}$$

3. Empirical analysis

3.1. Descriptive statistics

IOP Conf. Series: Earth and Environmental Science **77** (2017) 012016 doi:10.1088/1755-1315/77/1/012016

The descriptive statistics are as follows (see table 1):

- The average whole price of litchi is 19.10 CNY/kg in Beijing, 20.43 CNY / kg in Chengdu, 18.40 CNY/kg in Guangzhou, 16.28 CNY/kg in Hefei, 14.98 CNY / kg in Jiaxing, 17.81 CNY /kg in Nanjing, 20.92 CNY/kg in Shanghai, 13.51 CNY/kg in Shenyang, 17.66 CNY/kg in Changsha, 15.54 CNY/kg in Zhengzhou, 17.46 CNY/kg in Chongqing.
- As for litchi's highest price, it is 70.00 CNY/kg in Beijing, 60.00 CNY/kg in Chengdu, 40.00 CNY/kg in Guangzhou, 60.00 CNY/kg in Hefei, 66.00 CNY/kg in Jiaxing, 61.40 CNY/kg in Nanjing, 80.00 CNY/kg in Shanghai, 30.00 CNY/kg in Shenyang, 60.00 CNY/kg in Changsha, 50.00 CNY/kg in Zhengzhou, 60.00 CNY/kg in Chongqing.
- In terms of litchi's lowest price, it is 7.96 CNY/kg in Beijing, 3.80 CNY/kg in Chengdu, 5.00 CNY/kg in Guangzhou, 7.80 CNY/kg in Hefei, 7.00 CNY/kg in Jiaxing, 6.00 CNY/kg in Nanjing, 8.00 CNY/kg in Shanghai, 6.40 CNY/kg in Shenyang, 6.00 CNY/kg in Changsha, 8.00 CNY/kg in Zhengzhou, 2.60 CNY/kg in Chongqing.

Table 1. Descriptive statistics of litchi price.

Sales Market	Observations	Mean	Standard Deviation	Minimum	Maximum
Beijing	517	19.10	8.57	7.96	70
Chengdu	306	20.43	10.38	3.80	60
Guangzhou	735	18.40	6.75	5.00	40
Hefei	314	16.28	7.36	7.80	60
Jiaxing	550	14.98	5.96	7.00	66
Nanjing	298	17.81	10.13	6.00	61.4
Shanghai	633	20.92	10.69	8.00	80
Shenyang	247	13.51	3.96	6.40	30
Changsha	781	17.66	9.30	6.00	60
Zhengzhou	354	15.54	7.82	8.00	50
Chongqing	460	17.46	7.84	2.60	60

3.2. BP filter method

BP filter [28] requires the sample size to be large and the sample data be covariance stationary process or unit root process when the full sample length asymmetric filter is selected. Therefore, the sample data should be tested for stationary before BP filter. In this article, we use Eviews8.0 to test the stability of litchi wholesale price in China. The test results are shown in table 2: The original sequences of the litchi wholesale prices in Hefei, Chengdu, Guangzhou, Jiaxing, Nanjing, Shanghai, Shenyang, Changsha, Zhengzhou and Chongqing are stationary, while the original data in Hefei are not stationary sequence but conform to the requirement of the whole sample length asymmetric filtering after the first order difference. The original sequences of the wholesale price of litchi in Beijing, Chengdu, Guangzhou, Jiaxing, Nanjing, Shanghai, Shenyang, Shenyang, Changsha, Zhengzhou and Chongqing is I (0), while in Hefei is I (1) which need to be adjusted before using the full sample asymmetric BP filter. The methods used for detrending mainly contain removing the average, removing the linear trend and adjusting the drift. Considering that there may be mean value of the original data, this paper chooses the trend method removing the average to adjust.

Applying cycle factor based on BP filter method and "Crest-Crest" method, this paper divides wholesale price fluctuation of litchi into some cycles, as shown in table 3: from April 1, 2012 to September 30, 2016, litchi wholesale price in Beijing can be divided into 12 cycles; Chengdu into 10 cycles; Guangzhou into 23 cycles; Hefei into 10 cycles; Jiaxing into 15 cycles; Nanjing into 10 cycles;

Shanghai into 15 cycles; Shenyang into 10 cycles; Changsha into 21 cycles; Zhengzhou into 5 cycles; Chongqing into 10 cycles.

Table 2. Stability Test of litchi price in China.

Sales Market	Test form	ADF statistics	P value	ADF conclusion	KPSS statistics	1% critical value	KPSS conclusion
Beijing	C(C,T,0)	-6.4148	0.0000	Y	0.0855	0.7390	Y
Chengdu	C(C,T,0)	-6.3110	0.0000	Y	0.0563	0.7390	Y
Guangzhou	C(C,T,0)	-6.0026	0.0000	Y	0.2900	0.7390	Y
Hefei	C(C,T,0)	-6.6364	0.0000	Y	0.8897	0.7390	N
Jiaxing	C(C,T,2)	-5.0810	0.0000	Y	0.0860	0.7390	Y
Nanjing	C(C,T,0)	-5.2688	0.0000	Y	0.1266	0.7390	Y
Shanghai	C(C,T,0)	-7.7335	0.0000	Y	0.4526	0.7390	Y
Shenyang	C(C,T,0)	-7.5403	0.0000	Y	0.0735	0.7390	Y
Changsha	C(C,T,0)	-6.9880	0.0000	Y	0.1022	0.7390	Y
Zhengzhou	C(C,T,0)	-3.8623	0.0026	Y	0.1229	0.7390	Y
Chongqing	C(C,T,0)	-7.0794	0.0000	Y	0.1160	0.7390	Y
DHefei	C(C,T,0)	-15.2062	0.0000	Y	0.0221	0.7390	Y

D represents the first order difference for the time series.
In the test form (C, T, L), C is the intercept term, T is the trend term, L is the lag order, Y means stationary, and N means non-stationary.
The lag order is measured by the AIC criterion and the SIC criterion, and the maximum lag order applies the default setting.

3.3. *H-P filter method*

The main idea of H-P filter is to decompose the original time series into trend components and cyclic components [29]. After decomposing, we can get the H-P filter results of litchi wholesale price in China as follows:

Applying cycle factor based on H-P filter method and "Crest-Crest" method, we can divide the wholesale price fluctuation of Chinese litchi into different cycles according to different sales marktes, as shown in table 3: from April 1, 2012 to September 30, 2016, litchi wholesale price in Beijing can be divided into 12 cycles; Chengdu into 8 cycles; Guangzhou into 23 cycles; Hefei into 8 cycles; Jiaxing into 15 cycles; Nanjing into 10 cycles; Shanghai into 15 cycles; Shenyang into 8 cycles; Changsha into 21 cycles; Zhengzhou into 5 cycles; Chongqing into 13 cycles.

Table 3. Cycle comparison between BP filter and H-P filter.

Sales markets	BP filter	H-P filter	Sales Market	BP filter	H-P filter
Beijing	12	12	Shanghai	15	15
Chengdu	10	8	Shenyang	10	8
Guangzhou	23	23	Changsha	21	21
Hefei	10	10	Zhengzhou	5	5
Jiaxing	15	15	Chongqing	10	13
Nanjing	10	10			

After Comparing the filtering results of BP filter method and H-P filter method of the 11 sales markets below (see tables 4-14), we can find the cycle divided by BP and H-P are almost the same except in Chengdu and Shenyang whose cycles divided by BP filter are a little higher than that of H-P filter, Chongqing whose cycles divided by BP filter are a little less than that of H-P filter.

Table 4. Cyclical characteristics of price fluctuations in Beijing.

Cycle #	Wavelength (day)		Pitch		The position of the trough in the wavelength	
	BP	H-P	BP	H-P	BP	H-P
Cycle 1	7	20	4	10	6	8
Cycle 2	8	30	2	44	3	18
Cycle 3	9	50	1	46	4	40
Cycle 4	8	50	0.5	20	4	30
Cycle 5	7	10	2	12	5	6
Cycle 6	6	30	1	14	4	16
Cycle 7	10	40	6	2	3	21
Cycle 8	8	80	2	6	4	10
Cycle 9	9	20	10	8	5	12
Cycle 10	7	40	3	12	4	22
Cycle 11	12	40	35	14	4	22
Cycle 12	10	50	11	12	6	40

Table 5. Cyclical characteristics of price fluctuations in Chengdu.

Cycle #	Wavelength (day)		Pitch		The position of the trough in the wavelength	
	BP	H-P	BP	H-P	BP	H-P
Cycle 1	20	25	15	4	5	15
Cycle 2	21	50	8	16	11	40
Cycle 3	43	15	18	5	37	5
Cycle 4	19	50	25	26	12	40
Cycle 5	47	20	39	7	40	6
Cycle 6	18	30	27	26	14	17
Cycle 7	35	28	33	15	25	15
Cycle 8	25	40	34	20	19	21
Cycle 9	36		30		31	
Cycle 10	19		31		15	

Table 6. Cyclical characteristics of price fluctuations in Guangzhou.

Cycle #	Wavelength (day)		Pitch		The position of the trough in the wavelength	
	BP	H-P	BP	H-P	BP	H-P
Cycle 1	15	22	5	4	11	10

Cycle 2	19	34	3	26	15	12
Cycle 3	25	39	21	12	5	19
Cycle 4	17	32	7	16	13	15
Cycle 5	34	40	11	18	6	24
Cycle 6	30	48	13	32	22	23
Cycle 7	39	20	15	20	19	8
Cycle 8	45	45	25	25	23	35
Cycle 9	20	21	17	15	15	12
Cycle 10	40	9	27	2	31	5
Cycle 11	22	28	13	21	16	17
Cycle 12	33	16	10	10	31	10
Cycle 13	23	14	6	3	2	7
Cycle 14	24	15	7	4	8	8
Cycle 15	22	39	24	33	10	14
Cycle 16	33	23	6	7	10	10
Cycle 17	47	31	16	29	11	23
Cycle 18	43	18	28	2	39	9
Cycle 19	40	17	27	1	36	8
Cycle 20	20	31	2	20	2	2
Cycle 21	24	50	6	16	22	24
Cycle 22	35	16	17	8	10	10
Cycle 23	39	62	4	24	35	41

Table 7. Cyclical characteristics of price fluctuations in Hefei.

Cycle #	Wavelength (day)		Pitch		The position of the trough in the wavelength	
	BP	H-P	BP	H-P	BP	H-P
Cycle 1	29	18	5	3	6	10
Cycle 2	64	36	25	7	9	16
Cycle 3	13	28	19	2	8	17
Cycle 4	23	32	13	15	10	18
Cycle 5	32	30	12	11	9	20
Cycle 6	19	32	9	14	10	20
Cycle 7	17	26	8	6	8	22
Cycle 8	15	28	6	5	7	23
Cycle 9	19	20	23	15	15	12
Cycle 10	33	32	20	14	28	18

Table 8. Cyclical characteristics of price fluctuations in Jiaxing.

Cycle #	Wavelength (day)		Pitch		The position of the trough in the wavelength	
	BP	H-P	BP	H-P	BP	H-P
Cycle 1	24	18	7	4	15	8
Cycle 2	19	20	4	8	8	9
Cycle 3	21	36	9	16	14	16
Cycle 4	39	30	11	20	13	12
Cycle 5	30	24	16	8	6	18
Cycle 6	27	50	10	26	17	15
Cycle 7	53	22	21	20	20	20
Cycle 8	22	42	18	6	12	19
Cycle 9	41	30	4	7	18	16
Cycle 10	30	28	6	4	15	18
Cycle 11	29	32	3	8	17	18
Cycle 12	30	22	6	2	20	11
Cycle 13	50	25	23	17	30	12
Cycle 14	58	55	17	16	23	21
Cycle 15	31	34	11	6	13	14

Table 9. Cyclical characteristics of price fluctuations in Nanjing.

Cycle #	Wavelength (day)		Pitch		The position of the trough in the wavelength	
	BP	H-P	BP	H-P	BP	H-P
Cycle 1	25	32	20	14	13	18
Cycle 2	9	22	6	15	5	16
Cycle 3	40	18	16	4	33	10
Cycle 4	27	10	8	12	12	5
Cycle 5	50	20	14	5	41	14
Cycle 6	25	25	28	12	21	20
Cycle 7	10	30	17	7	6	18
Cycle 8	33	34	16	29	7	14
Cycle 9	10	40	18	22	7	22
Cycle 10	11	30	4	12	8	18

Table 10. Cyclical characteristics of price fluctuations in Shanghai.

Cycle #	Wavelength (day)		Pitch		The position of the trough in the wavelength	
	BP	H-P	BP	H-P	BP	H-P
Cycle 1	39	40	24	23	23	15
Cycle 2	14	40	10	25	8	25

Cycle 3	35	28	20	8	13	12
Cycle 4	28	38	22	13	14	15
Cycle 5	132	56	23	7	18	40
Cycle 6	22	46	18	13	13	25
Cycle 7	30	26	6	12	21	14
Cycle 8	9	16	30	18	4	8
Cycle 9	22	18	38	12	14	8
Cycle 10	8	25	40	10	3	14
Cycle 11	57	28	21	17	40	18
Cycle 12	74	36	10	12	68	16
Cycle 13	32	76	28	20	26	60
Cycle 14	30	38	50	50	13	19
Cycle 15	21	20	32	28	17	8

Table 11. Cyclical characteristics of price fluctuations in Shenyang.

Cycle #	Wavelength (day)		Pitch		The position of the trough in the wavelength	
	BP	H-P	BP	H-P	BP	H-P
Cycle 1	20	18	3	2	5	8
Cycle 2	10	30	4	2	7	12
Cycle 3	20	28	3	9	5	10
Cycle 4	27	22	18	5	10	10
Cycle 5	21	23	11	4	16	8
Cycle 6	9	20	12	1	4	10
Cycle 7	10	36	5	8	5	22
Cycle 8	20	28	4	6	14	18
Cycle 9	32		5		5	
Cycle 10	30		8		20	

Table 12. Cyclical characteristics of price fluctuations in Changsha.

Cycle #	Wavelength (day)		Pitch		The position of the trough in the wavelength	
	BP	H-P	BP	H-P	BP	H-P
Cycle 1	18	18	7	6	15	12
Cycle 2	28	26	3	3	9	10
Cycle 3	31	34	18	18	16	7
Cycle 4	30	28	22	22	21	20
Cycle 5	27	24	5	1	9	9
Cycle 6	35	50	4	10	14	23
Cycle 7	81	72	7	24	41	35
Cycle 8	21	20	31	21	11	10

Cycle 9	24	32	25	15	18	18
Cycle 10	20	80	2	10	10	25
Cycle 11	65	50	11	6	31	38
Cycle 12	48	16	13	3	34	10
Cycle 13	16	40	10	7	11	34
Cycle 14	44	20	8	2	32	8
Cycle 15	42	32	5	4	28	15
Cycle 16	51	52	15	11	27	26
Cycle 17	13	18	13	5	10	10
Cycle 18	46	46	17	15	40	34
Cycle 19	22	20	27	19	16	9
Cycle 20	24	26	20	12	14	12
Cycle 21	25	30	19	15	23	22

Table 13. Cyclical characteristics of price fluctuations in Zhengzhou.

Cycle #	Wavelength (day)		Pitch		The position of the trough in the wavelength	
	BP	H-P	BP	H-P	BP	H-P
Cycle 1	57	50	7	45	7	32
Cycle 2	50	68	29	44	40	22
Cycle 3	11	38	28	22	8	26
Cycle 4	59	32	16	6	9	22
Cycle 5	145	72	13	20	60	5

Table 14. Cyclical characteristics of price fluctuations in Chongqing.

Cycle #	Wavelength (day)		Pitch		The position of the trough in the wavelength	
	BP	H-P	BP	H-P	BP	H-P
Cycle 1	11	42	12	12	9	25
Cycle 2	10	58	7	5	4	18
Cycle 3	51	44	14	10	46	31
Cycle 4	9	26	8	5	7	12
Cycle 5	26	40	11	8	15	32
Cycle 6	14	22	9	6	5	9
Cycle 7	35	30	18	9	31	23
Cycle 8	26	19	16	2	22	12
Cycle 9	10	21	11	7	7	13
Cycle 10	12	23	15	5	5	10
Cycle 11		29		9		17
Cycle 12		16		10		7
Cycle 13		20		13		14

According to the results above, we can get information about the wavelength, the pitch and the position of the trough in the wavelength of fluctuation of litchi wholesale price in China.

- In terms of wavelength: in Beijing, the longest wavelength obtained by BP filter method is 12 days falling in the eleventh cycle while 80 days by H-P filter, the shortest wavelength is only 6 days obtained by BP filter while 10 days by H-P filter; in Chengdu, the longest wavelength obtained by BP filter method is 43 days falling in the third cycle while 50 days by H-P filter, the shortest wavelength is 18 days falling in the sixth cycle while 15 days by H-P filter; in Guangzhou, the longest wavelength obtained by BP filter method is 45 days falling in the eighth cycle while 62 days by H-P filter and the shortest is only 9 days falling in the eighth cycle while 9 days by H-P filter; in Hefei, the longest wavelength obtained by BP filter method is 32 days falling in the second cycle while 32 days by H-P filter, the shortest wavelength is 15 days falling in the first cycle while 10 days by H-P filter; in Jiaxing, the longest wavelength obtained by BP filter method is 58 days falling in the fourth cycle while 55 days by H-P filter, the shortest wavelength is 19 days falling in the second cycle while 18 days by H-P filter; in Nanjing, the longest wavelength obtained by BP filter method is 50 days falling in the fifth cycle while 40 days by H-P filter, the shortest wavelength is 9 days falling in the second cycle while 10 days by H-P filter; in Shanghai, the longest wavelength obtained by BP filter method is 132 days falling in the fifth cycle while 76 days by H-P filter, the shortest wavelength is only 8 days falling in the eighth cycle while 16 days by H-P filter; in Shenyang, the longest wavelength obtained by BP filter method is 32 days falling in the ninth cycle, the shortest wavelength is 9 days falling in the sixth cycle; in Changsha, the longest wavelength obtained by BP filter method is 81 days falling in the seventh cycle, the shortest wavelength is 13 days falling in the seventeenth cycle; in Zhengzhou, the longest wavelength obtained by BP filter method is 145 days falling in the fifth cycle while 72 days by H-P filter, the shortest wavelength is only 11 days falling in the third cycle while 32 days by H-P filter; in Chongqing, the longest wavelength obtained by BP filter method is 35 days falling in the seventh cycle while 58 days by H-P filter, the shortest wavelength is 9 days falling in the fourth cycle while 16 days by H-P filter.
- Pitch shows the distance between the peak and the trough of litchi price, the lager the value of pitch is, the greater litchi price fluctuates. So as for pitch, in Beijing, the maximum value of the pitch obtained by BP filter method appears in the eleventh cycle indicating that litchi wholesale price in this period fluctuates largely, while in the eleventh cycle by H-P filter, the minimum value appears in the fourth cycle obtained by BP filter meaning that litchi wholesale price fluctuates largely in this period. while in the fourth cycle by H-P filter; in Chengdu, the maximum value of the pitch obtained by BP filter method appears in the fifth cycle indicating that litchi wholesale price in this period fluctuates largely, while in the fifth cycle by H-P filter, the minimum value appears in the second cycle obtained by BP filter meaning that litchi wholesale price fluctuates largely in this period, while in the second cycle by H-P filter; in Guangzhou, the maximum value of the pitch obtained by BP filter method appears in the eighteenth cycle indicating that litchi wholesale price in this period fluctuates largely, while in the eighth cycle by H-P filter, the minimum value appears in the twentieth cycle obtained by BP filter meaning that litchi wholesale price fluctuates largely in this period. while in the twentieth cycle by H-P filter; in Hefei, the maximum value of the pitch obtained by BP filter method appears in the second cycle indicating that litchi wholesale price in this period fluctuates largely, while in the second cycle by H-P filter, the minimum value appears in the first cycle obtained by BP filter meaning that litchi wholesale price fluctuates largely in this period. while in the first cycle by H-P filter; in Jiaxing, the maximum value of the pitch obtained by BP filter method appears in the seventh cycle indicating that litchi wholesale price in this period fluctuates largely, while in the seventh cycle by H-P filter, the minimum value appears in the eleventh cycle obtained by BP filter meaning that litchi wholesale price fluctuates largely in this period, while in the eleventh cycle by H-P filter; in Nanjing, the

maximum value of the pitch obtained by BP filter method appears in the sixth cycle indicating that litchi wholesale price in this period fluctuates largely, while in the sixth cycle by H-P filter, the minimum value appears in the tenth cycle obtained by BP filter meaning that litchi wholesale price fluctuates largely in this period, while in the tenth cycle by H-P filter; in Shanghai, the maximum value of the pitch obtained by BP filter method appears in the fourteenth cycle indicating that litchi wholesale price in this period fluctuates largely, while in the fourteenth cycle by H-P filter, the minimum value appears in the seventh cycle obtained by BP filter meaning that litchi wholesale price fluctuates largely in this period. while in the seventh cycle by H-P filter; in Shenyang, the maximum value of the pitch obtained by BP filter method appears in the fourth cycle indicating that litchi wholesale price in this period fluctuates largely, while in the fourth cycle by H-P filter, the minimum value appears in the second cycle obtained by BP filter meaning that litchi wholesale price fluctuates largely in this period, while in the second cycle by H-P filter; in Changsha, the maximum value of the pitch obtained by BP filter method appears in the eighth cycle indicating that litchi wholesale price in this period fluctuates largely, while in the eighth cycle by H-P filter, the minimum value appears in the second cycle obtained by BP filter meaning that litchi wholesale price fluctuates largely in this period. while in the second cycle by H-P filter; in Zhengzhou, the maximum value of the pitch obtained by BP filter method appears in the second cycle indicating that litchi wholesale price in this period fluctuates largely, while in the second cycle by H-P filter, the minimum value appears in the first cycle obtained by BP filter meaning that litchi wholesale price fluctuates largely in this period. while in the first cycle by H-P filter; in Chongqing, the maximum value of the pitch obtained by BP filter method appears in the ninth cycle indicating that litchi wholesale price in this period fluctuates largely, while in the ninth cycle by H-P filter, the minimum value appears in the second cycle obtained by BP filter meaning that litchi wholesale price fluctuates largely in this period, while in the second cycle by H-P filter.

- After calculating the location of the trough in the wavelength, we can find that the position of the trough in the fluctuation period of litchi wholesale price in China is not in the middle of this cycle. Usually, the wholesale price of Chinese litchi in the same fluctuation cycle falls faster from the first peak to the trough than the rate of price rising from the trough to the second peak of the descent cycle faster, which means Chinese litchi wholesale price fluctuates asymmetrically.

4. Conclusion

This paper analyzed the fluctuation characteristics of litchi wholesale price in China from April 1, 2012 to September 30, 2016 by BP filter and H-P filter. The main conclusions are as follows:

- The choice of the filter method has a weak influence on the division of the price fluctuation period;
- The fluctuations of wholesale price of litchi have strong cyclicality;
- The fluctuations of wholesale price of litchi are non-repeatable;
- The fluctuations of wholesale price of litchi are like "rocket and feather", which prove the asymmetric transmission theory of price.

4.1. Different filter methods weakly affect cycle divisions

The divisions of wholesale price cycle numbers of Beijing, Changsha, Guangzhou, Zhengzhou, Shanghai, Jiaxing, Nanjing and Hefei are nearly consistent between BP filter and H-P filter. The divisions of litchi wholesale price cycle numbers of Shenyang, Chongqing and Nanjing are not the same. The reasons for these differences may be the differences in the requirements of the different filter methods, or the existence of some unique features in litchi wholesale price fluctuations of Shenyang, Chongqing and Nanjing.

4.2. Litchi prices fluctuate cyclically

In the market of litchi, the wholesale prices of Shenyang, Beijing, Changsha, Zhengzhou, Guangzhou, Shanghai, Jiaxing, Nanjing, Hefei, Chongqing and Chengdu litchi have obvious cyclical fluctuations. The possible factors of cyclical fluctuations are mainly affected by external shocks, the specific performance in the following two aspects:

- Climatic factors. Because there are high temperature, high humidity, summer rain and winter dry subtropical monsoon climate in litchi planting areas,. To explain cycle characteristics of climatic factors, this paper takes the maximum temperatures and minimum temperatures (see figure 1) in 11 litchi sales markets from January 1 2012 to December 31, 2016 as an example.

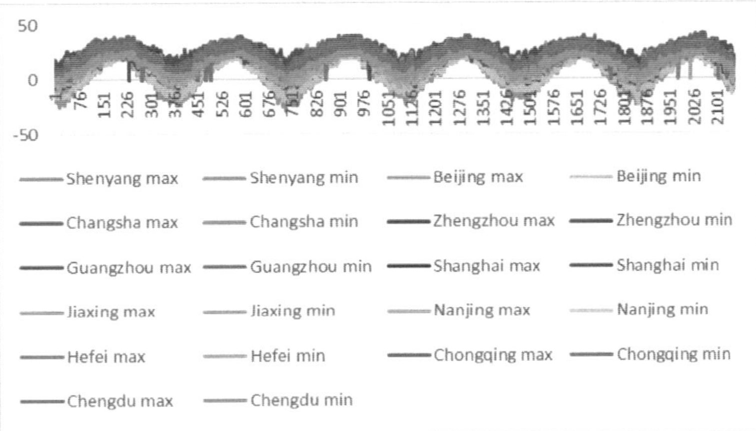

Figure 1. Maximum and minimum temperatures from 2012 to 2016.

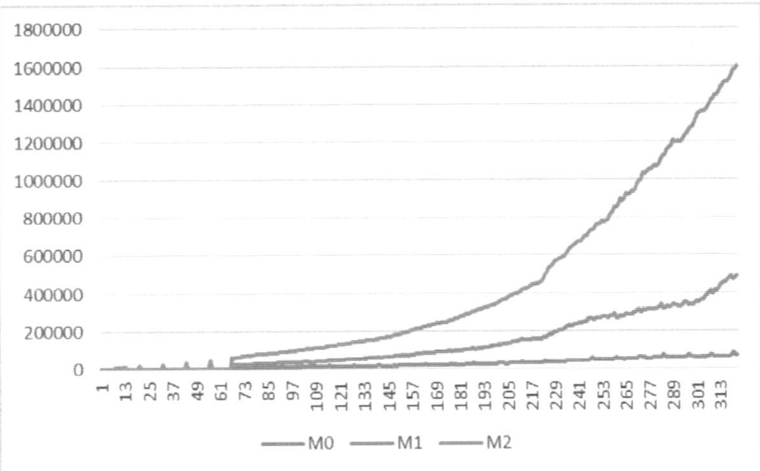

Figure 2. Money supply in China.

- Money supply. The theory of monetarism mainly concerns on the change of money supply to the price transmission. When the money supply changes, the price of the money is transferred to the non-agro-products, so that the overall price level of the whole society changes and is further transmitted to the agro-products. The impact of changes in the amount of money in short term more intense on prices of agro-products than non-agricultural products. The data released by the Bureau of Statistics show: from 2012 to 2016, the Chinese money supply experienced a continual augment (see figure 2, Unit: Billion RMB). According to the theory of monetarism, the change in the supply of money in the short term tends to have an impact on the prices of ago-products. In fact, the cyclical fluctuations of litchi price are caused not by the changes of money supply itself, but the percentage of money supply changes. The following analysis shows how the percentage change in money supply causes periodic fluctuations in litchi wholesale price.

Figure 3 shows: from 2012 to 2016, the percentage of change of money supply in China (Unit: %) shows cyclical fluctuations, these cyclical fluctuations transmit to the price of non-agro-products, and then the price fluctuations of non-agro-products transmit to the price fluctuations of agro-products. Cereals, vegetables, pork, mutton and other agro-products are necessity goods, and rational consumers would adjust their behavior in the consumption of agro-products based on their own preferences and budget constraints. On the one hand, the rural residents who are engaged in cultivation and breeding necessity goods are more elastic to needs than rural residents who are not. So, if the budget is constant, their consumption on necessities would not change much (mainly by self-sufficiency or small purchase), while the consumption of litchi changes just a little, this part of the litchi consumers are litchi price maintainers. On the other hand, the demand of rural residents who are not engaged in cultivating and breeding necessity goods and urban residents is more elastic than those who are engaged in cultivating and breeding necessity goods. Therefore, if the budget is constant, the shift of consumption of their necessities is relatively large (mainly by large purchases), and the consumption of litchi would be reduced. This part of the consumers is the causers of litchi price fluctuations.

To sum up, the cyclical fluctuations in the percentages of changes of money supply in China transmit to non-agro-commodity price fluctuations, and then the price fluctuations of non-agro-products transmit to the daily necessity agro-products, and at the end necessity agro-products cyclical price fluctuations transmit to litchi and other non-essential agro-products cyclical price fluctuations.

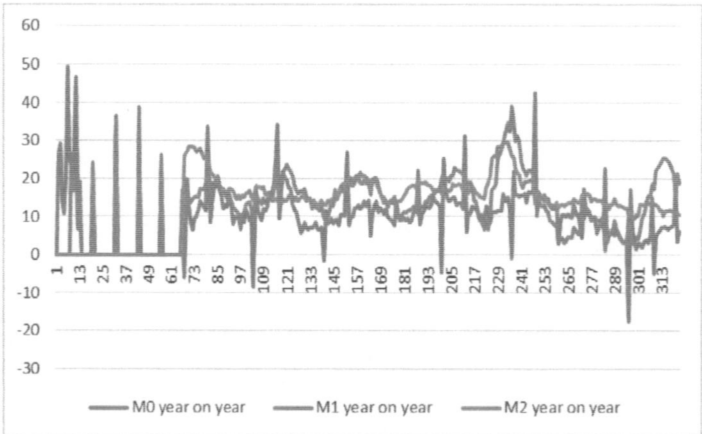

Figure 3. Percentage of change of China's money supply from 2011 to 2016.

4.3. Litchi wholesale price cyclical fluctuate non-repeatable

The fluctuation periods of litchi wholesale price in Shenyang, Beijing, Changsha, Zhengzhou, Guangzhou, Shanghai, Jiaxing, Nanjing, Hefei, Chongqing and Chengdu are not repeatable. The reasons for the cyclical non-repeatability of the wholesale price fluctuations are uncertain factors. The main uncertainties may be as follows:

- Natural disaster. Natural disaster, always sudden, which would cause some sudden change in litchi supply. For example, from July to August in 2012, frequent high temperatures continued in southern China; in late July "Vicente", "Sura", "Da Wei", "Swallow", "Kai Tak" and "Libra" Typhoon landed in sequence, and brought heavy rain. In the annual cycle of growth and development, litchi needs abundant water, but in the late development of fruits, heavy rain would deduce cracked fruit and produce fruit-drop. In 2013, low temperature in China's inland areas, the maturity of the producing areas across China was earlier than expected at the beginning of the year. Although litchi flowering was good, some of the producing areas suffered from continuous rainy weather during the flowering period. The fruits dropped severely, fruiting rate decreased. But due to appropriate application for technologies from Technology Demonstration Garden, flowering situation was good in most Demonstration Garden, and fruiting rate was higher than general level in local, causing an increase in litchi production in 2012, making the supply and quality of litchi improve over the previous year, and resulting in wholesale prices rising in varying degrees. In May 2014, Guangdong and Hainan were stricken by storms; "Madame", "Wilson", "Seagull" and "Phoenix" and other typhoons spread to Fujian and other origins, coincided with the fruiting period of early maturing litchi, so the fruits hit down by the storms. In October 25, 2015, Strong typhoon "Rainbow" landed Potou District in Zhanjiang, with big intensity, wide range, strongly destructive power, broke litchi branches and leaves; unfortunately, just as the flower-bud-differentiation period of early maturing litchi, thus produced lower fruiting rate. In 2016, "5.20" special heavy rain in Xinyi, Guangdong Province, caused a serious disaster, knocked off the fruits of early maturing litchi. In late October, "Hippocampus" landed in Guangdong, decreased the litchi fruiting rate. It can be seen that the types of abnormal climates, the time and place of occurrence, the frequency, the duration time, the different range of radiation, the degree of influence, and the rescue strategy of the grower cause changes of litchi supply. So the litchi price presents non-repeated cycle fluctuations. The wholesale price fluctuations from 2012 to 2016 were relatively frequent and uncertain, so some fluctuation cycle of litchi wholesale price is divided into several different small cycles.

- Concentrated litchi production. For the case of litchi, a kind of small agro-products, because of concentrated origins in four geographically adjacent provinces, namely Hainan, Guangdong, Guangxi and Fujian. Once the litchi origins suffered from natural disasters, there would be a sharp decline in production, so the litchi purchase price deviated from the expected fluctuation cycles, and then formed more deviation of the wholesale price fluctuation cycles than normal years through vertical transmission.
- Domestic high-quality litchi sales collectively. High quality domestic litchi sales together compressed importing quantity of litchi. In 2014, litchi sold together in Guangdong, Guangxi and Fujian, with good quality and taste, made Sino-Vietnamese fruit merchants undertake serious loss of importing litchi, and litchi imports from Vietnamese decreased. So, the domestic litchi total supply reduced, and demand for domestic litchi increased, leading to prices rising of litchi in China.
- Factors and paths are complex. The factors that affect the price fluctuation of litchi are numerous and the path of influence is complicated. Therefore, the periods of litchi wholesale price fluctuation may be close, but the factors that play a leading role will change with time. And new influencing factors will appear. Therefore, these two factors lead to the origins and sales markets of litchi price fluctuations to be non-repeatable.

4.4. Litchi wholesale price fluctuates like "rocket and feather"

- In Beijing, Chengdu, Guangzhou, Hefei, Jiaxing, Nanjing, Shanghai, Shenyang, Changsha, Zhengzhou and Chongqing, the wave pitches of litchi wholesale price fluctuations are asymmetric. The conclusion on "rocket and feather" asymmetric price transmission is consistent with the existing research. Litchi prices show "rockets and feathers" asymmetric fluctuations, and the main reasons are as follows: import foreign litchi. The price of domestic litchi is easily be impact on by importing foreign litchi. China imported litchi from Thailand, Malaysia, Vietnam and Myanmar in 2016. Among them, the export litchi from Vietnam to China was almost the same high quality as that of domestic litchi, but its orchards, packaging factories and warehouses must be registered in the Department of Agriculture in the origin country. The Pesticide content must go through the inspection and approval of agricultural administration system, and abide by the standard pesticide use management plan. In addition, packaging must comply with ISPM 15 standard, and PPD staff must conduct a rigorous inspection. As a result, Vietnamese litchi is favoured by Chinese consumers, even if their sales prices in China are much higher than that of domestic litchi, which leads to asymmetrical fluctuations in domestic litchi price.

- Alternative effect. The domestic litchis of major producing areas are mainly sold from April to May, and a small number of varieties in some producing areas continued to be sold in September. During the period, there are alternative products: mango and mangosteen in April; in strawberry in May; cherry in late May to mid-June; apricot in late May to mid-July; plum and peaches in July; watermelon in August; grape in September. As over the same period, alternative products are in abundance, along with diversified and changeable consumer preferences, so there is a hysteresis litchi between demand signal to send for consumers and consumption signal to receive for supply side, so price fluctuations are asymmetric.
- Bargain power. Bargain power of litchi suppliers, wholesalers and consumers in the supply chain are differences. On the one hand, litchi wholesalers' bargain power is lower than consumers. The wholesale price of litchi is mainly the result of the price game between litchi consumers and wholesalers (assuming there is only one wholesaler in the litchi wholesale market). And litchi wholesalers generally possess lower education than litchi consumers, so litchi wholesalers' bargain power lower than consumers in nature. On the other hand, the opportunity cost reduces the bargain power of litchi wholesalers. Because there is large number of alternative products during the same period, such as Vietnam litchi, watermelon, melon, etc. If litchi wholesalers are unwilling to sale litchi to consumers, then consumers will choose alternative products, considering the constraints of perishable traits of litchi, the opportunity cost to abandon the sale action is too big. So, to bear the profit reduction and the lower acquisition is more rational than to choose full loss. According to above two points, litchi consuming is increasing, and litchi demand curve shifts to the upper right, which makes the litchi wholesale price fluctuate.

Acknowledgments
This work is supported by the earmarked fund for China Agriculture Research System (Grand No. CARS-33-16).

References

[1] Lobos A G and Berbel J 2008 Volatilidad de precios internacionales recibidos por los productores de kiwis y manzanas frescas chilenas *Revista Brasileira De Fruticultura* **30** 133-9
[2] Gómez-Limón J A and Berbel J 2000 Multicriteria analysis of derived water demand functions: A Spanish case study *Agri.Syst.* **63** 49-72
[3] Griffith G R and Piggott N 1994 Asymmetry in beef, lamb and pork farm-retail price transmission in Australia *Agri. Econ. Agri. Economists* **10** 307-16
[4] Powers N J 1995 Sticky short-run prices and vertical pricing: Evidence from the market for iceberg lettuce *Agribusiness* **10** 57-75
[5] Goodwin B K and Harper D C 1999 Price transmission, threshold behavior, and asymmetric adjustment in the U.S. pork sector *J. Agri. Appl. Econ.* **32** 543-53
[6] Aguiar D R D and Santana J A 2002 Asymmetry and rigidity in farm to retail price transmission: Evidence from Brazil *Agribusiness* **18** 37-48
[7] Serra T and Goodwin B K 2010 Price transmission and asymmetric adjustment in the Spanish dairy sector *Applied Economics* **35** 1889-99
[8] Meyer J and Cramon-Taubadel S 2004 Asymmetric price transmission: A survey *J. Agri. Econ.* **55** 581-611

IOP Conf. Series: Earth and Environmental Science **77** (2017) 012016 doi:10.1088/1755-1315/77/1/012016

[9] Popovics P and Toth J 2006 Price transmission and asymmetric price development in the vertical structure of the milk industry in Hungary *Econ. Rev.* **53** 349-64
[10] Benkaabia M and Gil J M 2007 Asymmetric price transmission in the Spanish lamb sector *Europ. Rev. Agri. Econ.* **34** 53-80
[11] Jr O C and Sherwell P 2007 Alternative approaches in detecting asymmetry in farm-retail price transmission of fluid milk *Agribusiness* **23** 313-31
[12] Ioanna R and Yannis P 2008 Asymmetric price transmission in the Greek agri-food sector: Some tests *Agribusiness* **24** 16-30
[13] Ihle R and Amikuzuno J 2010 Assessing seasonal asymmetric price transmission in ghanaian tomato markets with the johansen estimation method *Courant Research Centre: Poverty, Equity and Growth - Discussion Papers* 53
[14] Xia T and Li X H 2010 Consumption inertia and asymmetric price transmission *J. Agri. Resour. Econ.* **35** 209-27
[15] Amikuzuno J and Ogundari K 2012 The contribution of agricultural economics to price transmission analysis and market policy in sub-Sahara Africa: what does the literature say? *Proc. Int. Conf. 86th on Agribusiness* (Coventry, UK: Warwick University)
[16] Nakajima T 2012 Asymmetric price transmission in the U.S. soybean exports *International J. Agri. Res.* **6** 368-76
[17] Loy J P, Holm T and Steinhagen C 2012 Vertical price transmission in differentiates product markets: A disaggregated study for milk and butter **89** 613-30
[18] Rezitis A N and Stavropoulos K S 2014 Price volatility and rational expectations in a sectoral framework model of the Greek meat market *Agri. Econ.* **42** 419-35
[19] Taghvaee V M and Hajiani P 2014 Price transmission mechanism in the Iran chicken market using the tecm, ecm-eg and gets models *Modern Economy* **5** 939-50
[20] Barahona J F, Trejos B, Lee J W, Chulaphan W and Jatuporn C 2014 Asymmetric price transmission in the livestock industry of Thailand *APCBEE Procedia* **8** 141-5
[21] Alam M J, McKenzie A M, Begum I A, Buysse J, Wailes E J and Huylenbroeck G V 2016 Asymmetry price transmission in the deregulated rice markets in Bangladesh: Asymmetric error correction model *Agribusiness* **32** 498-511
[22] Ilyasov J, Götz L, Kamiljon A, Dorosh P and Glauben T 2016 Market integration and price transmission in Tajikistan's wheat markets: Rising like rockets but falling like feathers? Ifpri Discussion Papers
[23] Harvey A C and Jaeger A 1998 Detrending, stylized facts and the business cycle *J. Appl. Econ.* **8** 609-25
[24] Cogley T and Nason J M 1995 Effects of the hodrick-prescott filter on trend and difference stationary time series implications for business cycle research *Real Busi. Cycles* 626-51
[25] Bunnoon P, Chalermyanont K and Limsakul C 2013 Multi-substation control central load area forecasting by using hp-filter and double neural networks (hp-dnns) International *J. Electric. Power Ener. Syst.* **44** 561-70
[26] He Y X, Wang B, Wang J H, Xiong W and Xia T 2013 Correlation between Chinese and international energy prices based on a hp filter and time difference analysis *Ener. Policy* **62** 898-909
[27] Xu W J, Gu R, Liu Y Z and Dai Y W 2015 Forecasting energy consumption using a new gm–arma model based on hp filter: The case of Guangdong province of China *Econ. Modelling* **45** 127-35
[28] Baxter M and King R G 1999 Measuring business cycles: approximate band-pass filters for economic time series *Rev. Econ. Statis.* **81** 575-93
[29] Hodrick R J and Prescott E C 1997 Postwar U.S. business cycles: An empirical investigation *J. Money Credit Banking* **29** 1-16

Planting structure adjustment and food security in major food production district: A case study on 10 main food production counties in Gansu Province, China

S J Wang[1, 3], C L Yang[2] and L Y Zhou[1]

[1] State Key Laboratory of Cryospheric Sciences, Northwest Institute of Eco-Environment and Resources, Chinese Academy of Sciences, Lanzhou 730000, Gansu, China
[2] Gansu Academy of Social Sciences, Lanzhou 730070, Gansu, China

E-mail: xiaohanjin@126.com

Abstract. This paper made an empirical study on planting structure adjustment and food security, in which main data and information came from the questionnaires of 10 main food-production counties in Gansu Province, China. The investigation results showed that: 1) During 1995 and 2014, the cultivated land area per household dropped by 2.40%, in which food crop area declined by 3.16%, yet cash crop area increased by 129% in the survey area. In the same period, the revenue per household increased by 162.99%, while food income from the revenue only increased by 17.42%; 2) In Hexi and Longzhong districts, mean wheat crop area per household shows a downtrend, while cash crop area increased significantly in the past 20 years. Especially, the household food output and income did not appear a simultaneous increase trend. In Longdong district, the household food output and income showed a simultaneous uptrend, and household income came mainly from miscellaneous grain (e.g. sorghum, bean, buckwheat, etc.); 3) In order to pursue higher economic efficiency of cultivated land, the farmers were forced to adjust planting structure and develop characteristics industries, profitable agriculture and cash crop with comparative advantage, which resulted in few food stock and impacted seriously on regional food security.

1. Introduction

In recent years, drought, flood, natural disasters, bio-fuels utilization etc., food prices continuously rose with extreme climate increasing. As a result, food crisis began to appear in the international market. Furthermore, some riots even happened in some countries. Currently, a threat of food insecurity was growing and food security has become a focus national government and economic theorists paid attention to [1-9]. In China, per capita arable land area is 40% less than that of the world average, food security has a great potential threat and has become one of the issues academic community focused on [10,11]. In 2013, food output reached 6.02×10^{11} kg, increased by 58.84% than 1985, and rose to its highest level since 1985. In 2003, food output (4.31×10^{11} kg) reduced to the lowest level since 1990. Therefore, the state began to carry out the policy of food subsidies in order to encourage farmers to expand sown are for food production. Since 2004, food production increased rapidly, far higher than the inter-annual level during 1985-2003 [12]. However, China's industrialization, urbanization and marketing process will accelerate in further and the cultivated land area will also continue to decline year after year, and a potential threat to food security will gradually emerge in recent years with the economic globalizing and the market internationalizing. In 2008, "National food security and long-term planning framework" stated that China's food self-sufficiency rate will stabilize at more than 95%, as of 2020, the amount of cultivated land will be maintained no less than 1.2×10^9 hm^2, the grain production capacity will reach 5.40×10^{12} kg, which fully showed national emphasis on food security.

Internationally, the strategic measures resolved food security is generally divided into three modes. The first is to achieve food self-sufficiency, the second is to use international trade to reach the balance goal of food supply and demand, and the third is to combine self-sufficiency and international trade for achieving food security. This paper agreed with the regional food supply and demand balance (self-sufficiency) mode, the main reason is that regional food security will seriously affect by extreme weather, industrialization, urbanization, some infectious diseases (e.g. Severe Acute Respiratory Syndromes) and the friendly relations between regions. When the supply and demand balance of regional food reach to self-sufficiency state, food security can be adequately safeguarded, otherwise, food safety will exist in a potential danger.

China's cultivated land resources, food production capacity and food supply-demand varies vastly in the different regions and provinces. Regional and sub-provincial food security is an integral part of national food security system, while food security of the major food production areas is the basis of regional food security, which should be received special attention. The major food production areas cannot only meet own regional food security and resist the threat of major natural disasters, but also maintain the dynamic equilibrium of food supply and demand inside and outside regions. Based on these reasons, this paper took three major food production areas of Gansu Province, China as the research object, analyzed change trend of crop production structure from 1980 to 2014, made a field survey and interviews faced to the farmers in ten major food production counties, investigated some factors and reasons planting structure adjustment affected food security, and brought forward some countermeasures and advice assorted with structural adjustment of crop production and raised the total grain output with a view to benefiting to stabilize China's food safety system and institute food production policy.

2. Study area

Gansu Province lies at the juncture of Qinghai-Tibet, Inner-Mongolia and Loess Plateaus in Northwest, China, and belongs to western inland arid areas (figure 1). Gansu has large land resources, but little precipitation, large evaporation and limited water resources lead to the lower percentage and capacity of utilizable land and cultivated land. Based on the above condition, Gansu`s food production mostly focused on wheat crop, and inter-regional food security mainly refers to wheat supply and demand balance. Since the 1990s, Gansu achieved basically an initial balance of food supply and demand with low level, and showed a coexistence phenomenon on the overall balance of supply and demand as well as the structural surplus and shortage [13, 14]. Currently, Gansu is still at a low level of food security, the balance of food supply and demand is not still very stable. In recent years, the depressed food prices and decreased farmer's income resulted in an ironic phenomenon that the benefit to plant grass is more profitable than that to plant wheat, all which resulted in the farmers adjusting dramatically crop production structure and reducing continuously food sown area.

In Gansu, the total sown area showed an insignificant increase trend (0.01×10^6 hm^2/a, $P<0.01$) during 1993-2014, but it increased rapidly after 2006. Food crop area decreased significantly with a rate of 0.07×10^6 hm^2/a ($P<0.0001$) in the past 20 years, but its output increased obviously with a rate of 0.16×10^9 kg/a ($P<0.0001$) in the same period. By contrast, cash crop area exhibited a significantly increase with a rate of 0.01×10^6 hm^2/a ($P<0.001$) in the same period (figure 2).

IOP Conf. Series: Earth and Environmental Science **77** (2017) 012017 doi:10.1088/1755-1315/77/1/012017

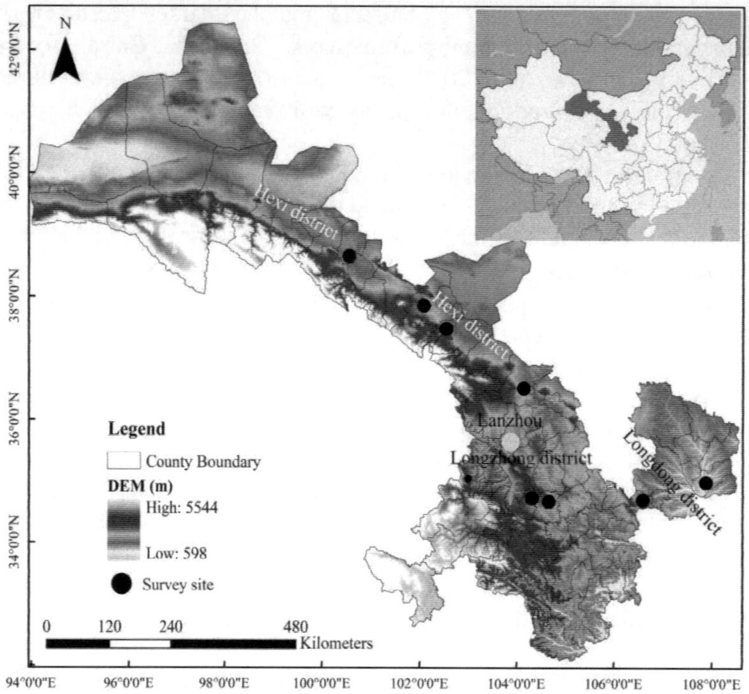

Figure 1. Location and distribution of the study area.

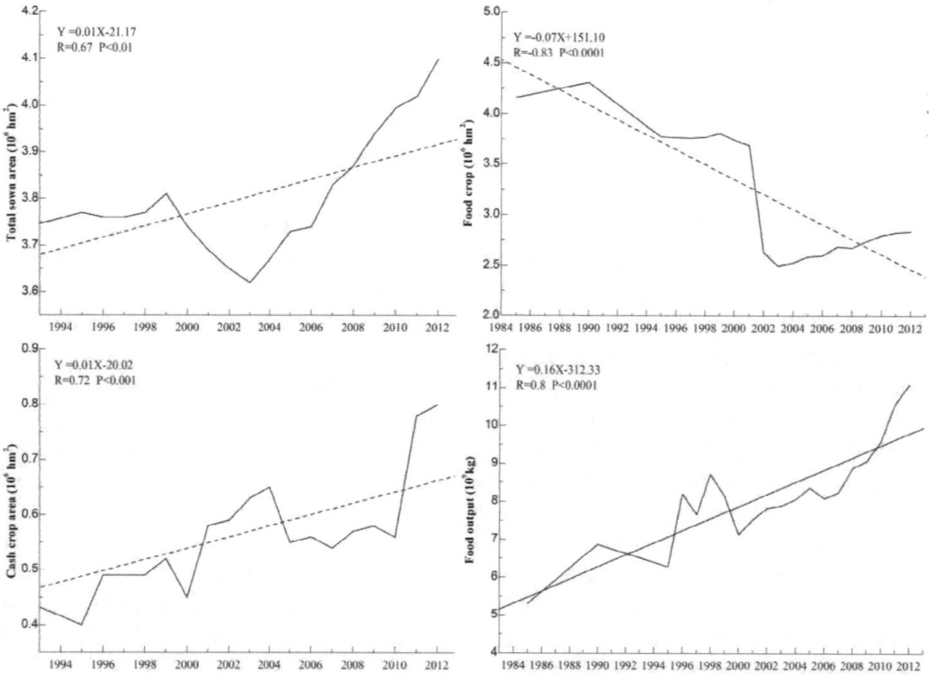

Figure 2. Change trend of crop sown area and food crop output in Gansu Province in 1985-2012. (Source: Gansu Statistical Yearbook and Rural Economics Yearbook, 1986-2013)

Of food crop area, the sown area of wheat showed a significant decrease ($P<0.001$) from 1985 to 2012, but the sown area of corn increased significantly by 7.63% ($P<0.001$). The decreasing part of wheat sown area, on the one hand, shifted corn area, on the other hand, turned into cash crops and other crop area, in which the sown area of soybean occupied small percentage in food crop area and remained about 3.0% (table 1). In 2004, Gansu province implemented entirely nation policies to exempt taxes of special agricultural products and increase subsidies for food, high quality seed and farm machinery. These policies got really farmers more benefits, and mobilized their enthusiasm to grow food crop, especially wheat. After 2004, the total sown area slightly increased, whereas food output increased significantly. By 2012, food output increased to 11.39×10^6 kg, rose by 65.89%, compared with 1990 (Figure 2(a)). However, with population increasing and industry and fodder food (especially wheat) consumption increasing, at present, Gansu has appeared a phenomenon on food supply less than demand in 8 years and food security problems gradually emerged.

Table 1. Sown area change of the main food crop in Gansu in 1985-2012.

Unit: 1×10^6 hm^2

Year	food crop	wheat	corn	soybean	other
1985	4.16	1.49	0.28	0.05	0.96
1990	4.31	1.46	0.30	0.06	1.06
1995	3.77	1.36	0.35	0.07	1.16
1999	3.76	1.22	0.53	0.09	1.07
2000	3.76	1.19	0.46	0.09	1.05
2001	3.77	1.12	0.47	0.08	1.02
2002	3.81	1.07	0.50	0.08	0.98
2003	3.74	0.96	0.49	0.08	0.97
2004	3.69	0.93	0.50	0.09	1.02
2005	2.63	0.96	0.50	0.09	1.03
2006	250	0.94	0.52	0.09	1.06
2007	2.53	0.92	0.50	0.10	1.17
2008	2.59	0.90	0.56	0.10	1.12
2009	2.60	0.92	0.70	0.09	1.03
2010	2.69	0.88	0.84	0.09	0.99
2011	2.68	0.86	0.84	0.08	1.02
2012	2.74	0.82	0.92	0.06	1.00

Source: Gansu Statistical Yearbook and Rural Economics Yearbook, 1986-2013.

3. Date source and method

The preliminary survey was carried out from July to August in 2014. For understanding the interrelationship among structure adjustment of food crop, food production and food security, the research team chose 500 farmers as a survey sample and spent two months to make a field research and direct visit to them from 10 counties in three main food production districts (i.e. Hexi, Longzhong and Longdong city) (figure 1). Finally, the research team issued and returned 300 survey questionnaires, consisting of 486 completed questionnaires and 14 uncompleted. The effective response rate reached 97%. The survey methods chose relative open form [15].

During the investigation, the team visited the village committee, the retired director, and some senior farmers, inquired them in detail on the development status of agricultural economy, structural adjustment of crop production and food security etc. Meanwhile, the research team surveyed some farmers in the form of the interview, the investigation contents included the basic farmers' situation that impacted food security, like crop production, crop area change, planting structure adjustment, the constitute of the household income and crop yield change during 20 years from 1995 to 2014. After investigating, the research team verified the survey contents with the village committee again and did some necessary complement. At the same time, the research team explained some data to be inconsistent with reality and made some error correction. Finally, the research team aggregated and processed the survey data by statistical software EXCEL, and analyzed some correlative problems on structural adjustment of crop production, annual household income, and food yield and food security in three major food production areas.

4. Results and analyses

4.1. An empirical study of Hexi district

Hexi district locates in the west of the Yellow River in Gansu, including Jiuquan, Zhangye, Wuwei, Jiayuguan and Jinchang city (figure 1). Hexi district, relatively developed area in the rural economy, is once one of China's 10 main food production bases, the rural population accounts for 16.46% of Gansu's population and per capita income ranks first in Gansu. The 180 rural families were surveyed in Wangjiabao village, Dangzhai town in Ganzhou district, Zhangye City, Fazhan village, Qingyuan town in Liangzhou district, Wuwei City, and Beihaizi village, Chengguan town in Yongchang County, Jinchang City in Gansu. The survey results showed that the cultivated land per household surveyed reached to 0.57 hm^2 in 2014, increased by 5.26%, compared with 1995, in which the sown area of food crop decreased significantly (R^2=0.91) with a rate of 0.14 hm^2/10a (P <0.001) during 1995-2014, yet cash crop showed a significant increase trend with a rate of 0.16 hm^2/10a (P <0.001) (R^2=0.94). Survey shows that 80% of respondents said all edible wheat is purchased by the market. These factors resulted in food output reduction and made it loses the overwhelming superiority status of main food product base in the past (figure 3).

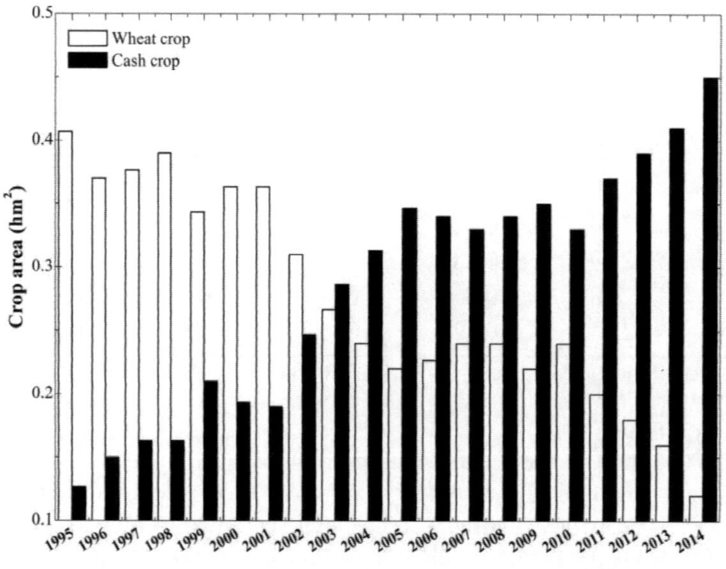

Figure 3. Crop production structure change of the households surveyed in Hexi district during 1995-2014.

The main reasons appeared these phenomena are as follows: over the past 10 years, national food policy took on a contraction status, and the comparative benefits of food industry decreased in most regions. In view of these reasons, the farmers changed the conventional behaviors to grow and reserve food, all which resulted in the sustaining downtrend of food planting area and food output per household. In addition, with cash crop (e.g. malting barley, wine grape planting, corn seed production, medicinal materials, vegetable in greenhouses and Medicago sativa, etc.) fast development relying on better soil and light-heat resources and temperature difference of day and night in Hexi corridor, wheat area downtrend will be inevitable under self-regulation influence of supply and demand balance in regional food market in the future.

In the previous period, Hexi district with 30% arable land of Gansu provided 70% commodity grain for Gansu province. However, food contribution rate decreased year by year, and now it has transferred to the main sales area from once the main producing area. At the same time, Hexi region has become the largest and most advantageous production base of corn seed in China. In 2013, the income per household investigated reached 38,000 RMB, increased 70.59%, while food income dropped to 1200 RMB, the reduction extent reached 40%, compared with 1995 (figure 4). Therefore, Hexi region should stabilize wheat crop area, develop food production, and implement innovation projects of agricultural science and technology in order to increase wheat production and ensure food stock goal and food supply and demand balance.

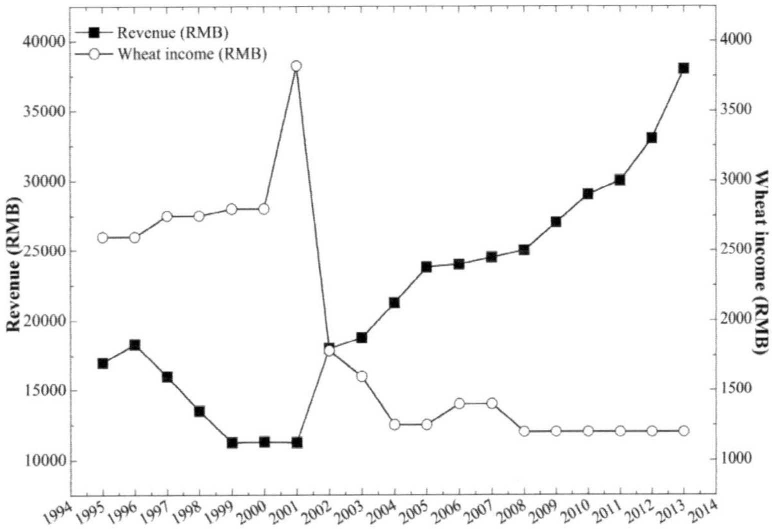

Figure 4. Revenue and wheat income per household change in Hexi district during 1995-2013.

4.2. An empirical study of Longzhong district

Longzhong locates in the abdominal region of Gansu, including Baiyin, Dingxi city and Linxia Huimin autonomous prefecture (figure 1), the rural population accounts for 34.19% of Gansu. Longzhong has larger dry land area and lower food production level, and is also a concentration region of the agricultural population. In Longzhong district, the research team surveyed 160 rural families of four villages, including Hanji village, Hanji town in Linxia Huimin autonomous prefecture, Linxia county, Fansiping village, Yujing town in Lintao county and Jiudianzi village, Pingxiang town in Tongwei county in Dingxi city, and Chenzhuang village, Xiquan town in Baiyin City, Jingtai county. During 20 years from 1995 to 2014, the cultivated land area per household surveyed appeared small uptrend. The mean area of wheat crop per household appeared a slight downtrend (0.06 hm^2/decade), especially, the downtrend increased clearly after 2000. The mean area of cash crop per household reached to 0.26 hm^2 in 2014, increased significantly by 420% ($P<0.001$), compared with 1995 (figure 5). At present, Longzhong has become the largest potato seed breeding base in Northwest, China.

During the period of 1995-2014, per household revenue showed a continuing upward trend (670.75 RMB/a) in Longzhong. Especially, per household revenue reached to 17000 RMB in 2014, and increased by 240%, compared with 1995. However, per household wheat income indicated a decrease with a rate of 276.5 RMB/decade ($P<0.001$, $R^2=0.55$) during the past 20 years. Especially, wheat income showed some increase trend after 2007 (figure 6). In contrast, income share from non-agricultural and cash crop appeared uptrend. As a whole, the agricultural structure increasingly changed towards the regional nature resources advantage and economic benefit. Similarly, Longzhong existed also a phenomenon that food output increased without income from wheat crop increasing.

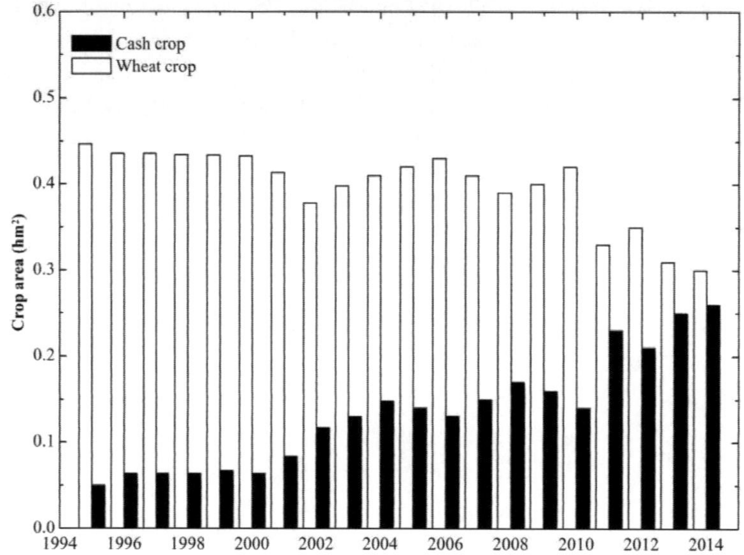

Figure 5. Crop production structure change of the households surveyed in Longzhong district during 1995-2014.

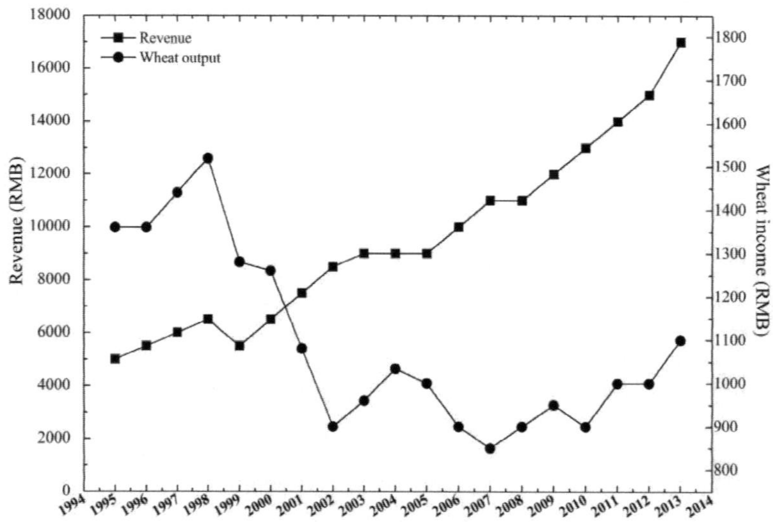

Figure 6. Revenue and wheat income per household change in Longzhong district during 1995-2013.

Based on these reasons, Longzhong should keep the basic crop area and exert to develop food production base. Longzhong should further adjust the proportion of summer and autumn crops and soybean in crop production, reduce appropriately the sown area of summer crops, increase moderately the sown area of autumn crops, particularly, and increase the sown area of high-yield corns, potatoes, soybeans, coarse cereals, ect. In the Yellow River coast used the electricity to irrigate, the most land is plain and the irrigable land also is relative concentrated, in view of this, the region should improve the food yield per unit by intensive cultivation, and built the commodity grain base in order to stabilize the supply and demand balance in inner region, Gansu.

4.3. An empirical study of Longdong district

Longdong district, locates in a typical region in Loess Plateau, including Pingliang and Qingyang city (figure 1), is one of the major food production bases in Gansu. Due to the geographical location and special nature conditions, Longdong takes on some obvious differences in crop production structure and farmers' behavior, compared with Hexi and Longdong districts. In Longdong, the research team surveyed 160 rural families in three villages, including Yuanquancheng village, Gaoping town and Gaohu village, Baima town in Qingcheng county, Pingliang city, and Ren village, Jiaocun town in Ning County, Qingyang city. The survey results showed that the cultivated area per household surveyed decreased by 20.74% in 2014, compared with 1995. Wheat area per household surveyed showed a first decrease and then increase trend in the past 20 years, while cash crop area per household showed a trend of first increase and then decrease (figure 7).

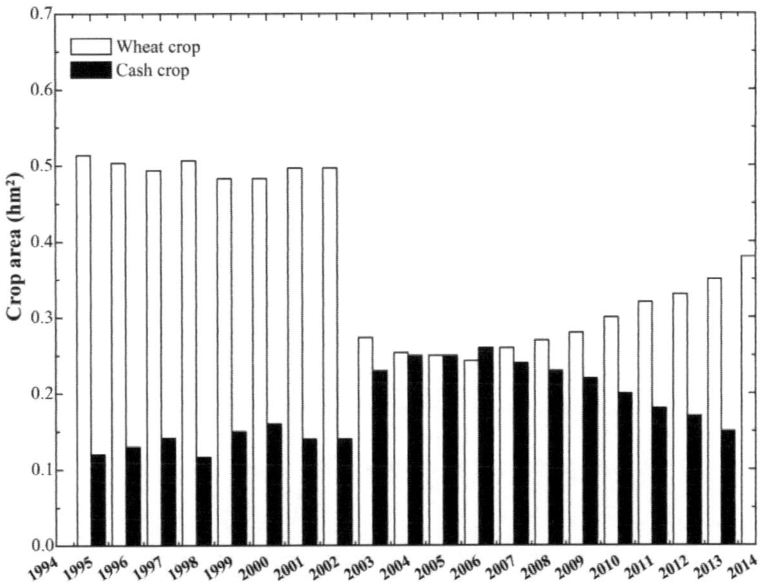

Figure 7. Crop production structure change of the households surveyed in Longdong district during 1995-2014.

The main reason was that the rehabilitating plough to forest (grassland) and pastureland rehabilitation project implemented on the large range in Gansu, the urbanization process accelerated further, the original sloped farmland turned into woodland and grassland, and some arable lands turned into urban areas, all which resulted in the cultivated land area decreasing. During 20 years from 1995 to 2014, the farmers' income came mainly from food crop. Although the sown area of food crop decreased to some extent, farmers' revenue and wheat income increased synchronously in the past 20 years. In the same period, the farmers' revenue and wheat income increased significantly with a rate of 589.47 RMB/a ($P < 0.001$, $R^2=0.98$) and 156.71 RMB/a ($P < 0.001$, $R^2=0.97$). Especially, in 2014, per household revenue and wheat income reached to respectively 16,000 RMB and 3, 900 RMB, rose by 214% and 195%, compared with 1995 (figure 8). However, food self-balance in Longdong now begins to appear inadequate.

However, food output and farmers' income in Longdong appeared the same change trend, in contrast to Hexi and Longzhong districts. The main reason was that Longdong was suitable for planting autumn crops and pollution-free vegetables because of special soil and weather conditions. In recent years, food output and market of miscellaneous grains, such as sorghum, bean, and buckwheat thrived especially. The farmers not only increased wheat output, but also effectively increased economic income and formed a benign interactive development pattern. In view of this, Longdong should prioritize to stabilize the basic farmland area, develop food production base, and mainly improved food yield per unit in order to increase food overall yield. Meanwhile, Longdong should also reduce moderately the summer crop area, enlarge the autumn crop area, increase properly cereal crop area, and extend the high yield corn crop area in the places with better water and heat conditions. In addition, Longdong should also spread cash crop and forage crop area, and develop vegetable and fruit industry in order to increase farmers' economic income.

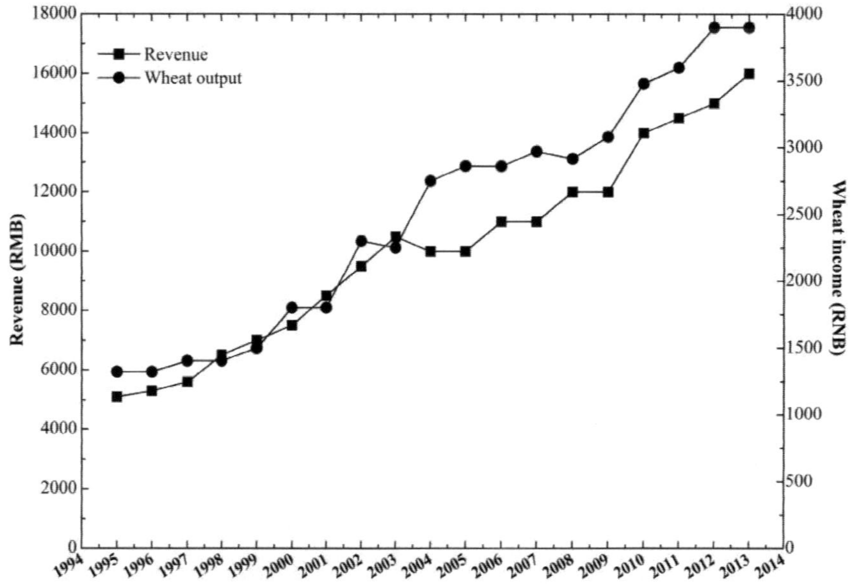

Figure 8. Revenue and wheat income per household change in Longdong district during 1995-2013.

4.4. Overall survey result

In the past 20 years, the cultivated land area of the surveyed farmers decreased by 2.40%, in which food crop area decreased by 3.16%, whereas yet cash crop area increased by 129%. At the same time, the surveyed farmer income per household increased 162.99%, while the income from food crop only increased by 17.42% during 1995-2013. The investigation data showed that Hexi and Longzhong districts appeared a trend that food output increased without income (from wheat) increasing after 2001. The main reasons are as follow: (1) The inputs cost of agriculture crop production increased continuously; (2) The income from wheat crop remained lower; (3) Gansu lacked for better quality agricultural products with higher prices. In order to pursue higher economic efficiency of cultivated land, the farmers were forced to adjust the agriculture structure and develop characteristics industries, profitable agriculture and cash crop growing with comparative advantage. In addition, with labor force moving and wheat output falling, most of farmers' food stock is very few and food stock is used only by them next year, all which impacted seriously on food security. With the policies on rehabilitating plough to forest (grassland) and pastureland rehabilitation project implementing on the large range and water-saving agriculture industry establishment, it is difficult to change entirely food output downtrend.

5. Conclusion

With the policies on rehabilitating plough to forest (grassland) and pastureland rehabilitation project implementing on the large range and water-saving agriculture industry establishment, it is difficult to change entirely food output downtrend. At the same time, the inputs cost of agriculture crop production increased continuously and the income from wheat crop remained lower in recent decades. State lacked for several better agricultural products with higher quality and prices. The farmers were forced to adjust the agriculture structure and develop characteristics industries, profitable agriculture and cash crop growing with comparative advantage in order to pursue higher economic efficiency of cultivated land, all which impacted seriously on state food security.

Therefore, state food security must rely solely on the major food production areas. The major food production areas must strictly protect cultivated land area, increase science and technology investment, improve the legal system of food security, and implement agriculture zoning program. Especially, considering global economic change, international security, domestic agricultural development, natural disasters, emergencies and inhabitants' consumer habits and other relevant factors, it is not realistic to depend highly on foreign food import. The major food production areas must pay attention to adjust and optimize crop production structure and increase food output, focus on food quality, and take it as an opportunity to improve farmers enthusiasm to grow food crop and increase farmers' income in order to achieve food supply and demand balance and food security in the main food production and sale areas.

Acknowledgments
This work was funded by National Natural Science Found (41690143, 41671058) and Technology Services Network Program of Cold and Arid Regions Environmental and Engineering Research Institute, Chinese Academy of Sciences (HHS-TSS-STS-1501).

References

[1] Anderson S 1990 Core indicators of nutritional state for difficult to sample populations *J. Nutr.* **102** 1559-660
[2] WHO 2002 *WHO Global Strategy for Food Safety: Safer Food for Better Health* (in English) (Geneva, Switzerland: World Health Organization)
[3] Liu X M 2004 *Research on China's Strategy of Food Security* (Beijing: China Market Press)
[4] Xiao G A 2005 *Research on China's Food Security* (Beijing: China Economic Publishing House) (in Chinese)
[5] FAO 2007 *The State of Food and Agriculture: Paying Farmers for Environmental Services* (Rome)
[6] Ericksen P J 2008 Conceptualizing, food systems for global environmental change research *Global Environ Chang* **18** 234-45
[7] Lu S Y and Sun W A 2008 Problems and countermeasure on food safety in China *Shandong Agric. Sci.* **7** 116-8 (in Chinese)
[8] Dong Y L, Yu J B and Liu Z X 2008 Agricultural trade liberalization, poverty and food security-the Indian experience *J South Chin. Agric. Univ. (Soc. Sci.)* **4** 21-8 (in Chinese)
[9] Marvin H J P, Kleter G A, Frewer L J, Cope S., Wentholt M T A and Rowe G 2009 A working procedure for identifying emerging food safety issues at an early stage: Implications for European and international risk management practices *Food Control* **20** 355
[10] Wang Y P 2005 Food Security and Protection and Sustainable Development (Beijing: China Agriculture Publishing House) (in Chinese)
[11] Zheng Y G 2007 Icebreaking policies and theoretical innovation concerning the "peasants, agriculture, and rural areas" *Teaching Res.* **9** 10-8
[12] National Bureau of Statistics of China 2007 *China Statistical Yearbook for Agriculture* (Beijing: China Statistics Press) (in Chinese)
[13] Chen J S and Yu X 2004 China's rural economic analysis of the situation in 2003 and outlook in 2004 *Chin. Rur. Eco.* **2** 23-9 (in Chinese)
[14] Chen Q Q and Li X W 2005 Thought on food security and income of the farmers in Gansu Province *J. Gansu Agric. Univ.* **40(1)** 117-21 (in Chinese)
[15] MaCracken J A, Pretty J N and Conway G R 1998 *An Introduction to Rapid Rural Appraisal for Agricultural Development* (London: International Institute for Environment and Development)

IOP Conf. Series: Earth and Environmental Science **77** (2017) 012018 doi:10.1088/1755-1315/77/1/012018

Development of newly enriched bread with quinoa flour and whey

D M Salazar[1, 3], M Naranjo[1], L V Pérez[1], AF Valencia[1], L P Acurio[1], L M Gallegos[2], F C Alvarez[1], P I Amancha[1], M P Valencia[2], C A Rodriguez[1] and M Y Arancibia[1]

[1] Department of Food Science and Engineering, Technical University of Ambato. Av. Los Chasquis & Río Payamino, Z.C. 180150, Ambato
[2] Department of Health Science, Technical University of Ambato. Colombia St. & Chile, Z.C. 180103, Ambato

E-mail: dm.salazar@uta.edu.ec

Abstract. Ecuador, Bolivia, and Peru are countries with the highest amount of quinoa production in the world due to the proximity to the Andes. Further, Ecuador has a high production of dairy products, particularly fresh cheese of which production gives a high volume of whey, without further use, with the consequent loss of their nutritional value. The present study was performed to develop a new fortified bread through the incorporation of quinoa flour and whey at three different concentrations. The use of quinoa and whey improved the texture, shelf life and sensory characteristics of bread, compared to those prepared with wheat flour. This study shows the potential of quinoa flour and whey as ingredients in the development of baked products.

1. Introduction

Bread has been regarded for centuries as one of the most popular staple food products [1]. Epidemiological studies suggest that people diet significantly prevents many chronic diseases [2]. Since bread is a staple food in developing countries, it is considered to be the best way for fortification [3]. However, from 1999 to 2003, the consumption of bread declined by 13.8%, due to such factors as changing nutritional habits and increasing use of substitutes e.g. breakfast cereals and fast foods [4]. At this time, consumers are demanding healthy diets and nutritional foods, these requirements have allowed a continuous growth on sales of functional foods and in the process establishing an enormous commercial market [1]. Functional foods are defined as innovative, physiologically-active foods, which can provide additional health benefits beyond basic nutrition. Food fortification is one of the major techniques used to create functional food products [5].

New varieties of baked goods are becoming increasingly popular worldwide. Generally, from a consumer's point of view, bread should have a low glycemic index, be a source of protein and dietary fiber, vitamins, magnesium, calcium, microelements, and antioxidants [6]. Necessary components for bread production are cereal flour, water, yeast, and salt, but some optional ingredients can be added to improve processing or to produce specialty and novelty bread which often have an increased nutritive value. Quinoa is a pseudocereal and is a native plant in the Andean region. Quinoa grains are considered as potentially gluten-free with an excellent nutrient profile [2, 7]. They contain considerable amounts of fiber and minerals, such as calcium and iron [2, 8]. It is also rich in antioxidants such as polyphenols [3, 9, 10]. Dairy ingredients are widely used in the baking industry because of their beneficial effect on the nutritional, organoleptic and some functional properties [11, 12]. Whey is defined as the fraction of milk obtained by rennet action or by acids during cheese making. It constitutes 90% of milk and contains water-soluble compounds. In this solution, it is possible to find soluble proteins, lactose, vitamins, and minerals [13]. Whey proteins represent about 20% of milk proteins and constitute a rich source of complete and bio-available amino acids. Lysine (9.1 g/100 g of whey protein) is a crucial amino acid that is lacking in a diet limited to direct consumption of cereal grains (1.6 g/100 g of wheat protein) [14,15]. The aim of this research was to establish the effect of the inclusion of quinoa flour and whey in bread.

2. Materials and methods

2.1. Materials

Whey was obtained from fresh cheese processing in a local factory. Pasteurization was performed at 63°C for 30 minutes. Fresh yeast (Levapan S.A, Ecuador) (4 g/100 g of flour), salt (Ecuasal S.A. 2 g/100 g of flour). Butter (10 g/100 g of flour), sugar (9 g/100 g of flour), eggs (9 g/100 g of flour), quinoa and commercial wheat flours were purchased from a local supermarket.

2.2. Breadmaking process

The quinoa flour and whey were added at three substitution levels: 10%, 15% and 20% (w/w) and 50%, 75% and 100% (w/w), respectively. The control sample was prepared with wheat flour and water. Whey was poured directly to the dough. Fresh yeast and salt were diluted in water before being added to the flour. The amount of added water for the control samples was 60 g per 100 g of flour. Water for the experimental bread was calculated according to the levels of substitution (table 1). All ingredients were mixed for 3 min., in a mixer (Ferton, USA). The dough was divided into portions of 180 g and proofed at 37°C and 80% relative humidity (RH) for 60 min. The dough was then baked in a rotatory oven (Zucchelli Mini Fanton, USA) at 210°C for 20 min. For each formulation, the bread samples were analyzed in triplicates obtained from different bread loaves.

Table 1. Formulation for bread with substitution of quinoa flour and whey.

Samples	**Wheat flour (g)**	**Quinoa flour (g)**	**Water (g)**	**Whey (g)**
Control	100	-	60	-
BQW501	90	10	30	30
BQW502	90	10	15	45
BQW503	90	10	-	60
BQW751	85	15	30	30
BQW752	85	15	15	45
BQW753	85	15	-	60
BQW101	80	20	30	30
BQW102	80	20	15	45
BQW103	80	20	-	60

BQW501: bread from quinoa flour 10% and whey 50%, BQW502: bread from quinoa flour 10% and whey 75%, BQW503: bread from quinoa flour 10% and whey 100%, BQW751: bread from quinoa flour 15% and whey 50%, BQW752: bread from quinoa flour 15% and whey 75%, BQW753: bread from quinoa flour 15% and whey 100%, BQW101: bread from quinoa flour 20% and whey 50%, BQW102: bread from quinoa flour 20% and whey 75%, BQW103: bread from quinoa flour 20% and whey 100%.

2.3. Technological characteristics

IOP Conf. Series: Earth and Environmental Science **77** (2017) 012018 doi:10.1088/1755-1315/77/1/012018

Height measurements were performed from the bottom to the highest point of the top of the bread. The texture analysis was carried out using a texturometer (CT3 Brookfield texture analyzer, USA). Probe speed was set to 2 mm/s to compress the center of the bread. Hardness values, corresponding to the maximum peak force of the first compression were recorded.

The slicing process was performed using an automatic cutter until obtain 10 mm thickness slices. The slicing was evaluated by weight loss before and after cutting and applying the equation (1).

$$Slicing\ (\%) = \frac{100 \times (initial\ weight - final\ weight)}{initial\ weight} \quad (1)$$

2.4. Physicochemical properties
The pH was determined using a potentiometer (HI 9126 HANNA, Rhode Island, USA). The acidity was determined by titration with 0.1 N NaOH, using phenolphthalein as indicator (AOAC 947.05). Total solids analysis was performed following the standard method NTE INEN 95. The moisture of bread was measured according to AOAC standard method (AOAC, 2005).

2.5. Colour determination
After the bread was made, the colour parameters were measured, L *(Brightness), a* (red/green) b*(yellow / blue), yellow index (YI) and brightness (B) with a Hunter Lab Colorimeter (mini Scan 4500L EZ, Hunter Associates Laboratory INC, Reston, Virginia, USA) calibrated with an illuminator D65 (natural light) and standard observer D10. The chroma polar coordinate or saturation C* was calculated from the expression C* = √ (a* 2 + b* 2). At least 15 measurements were performed in different areas of the sample, and the average recorded as the reported value.

2.6. Microbiological analysis and shelf life
Total aerobic mesophilic bacteria were evaluated following the official method AOAC 990.12. Molds and yeasts analyses were performed following the AOAC official method 997.02. The shelf life of baking products was determined based on total aerobic growth after a week period and the calculation of cfu/g of bread. The results were linearized to obtain the regression equation.

2.7. Proximal analysis
Fat was evaluated following the official method AOAC Ed.19 2012 2003.06; moisture following the official method AOAC Ed.19 2012 925.10; protein following the official method AOAC Ed.19 2001.11 and ash following the official method AOAC Ed.19 2012 923.03. This analysis was performed only for the best treatment.

2.8. Sensory evaluation
The sensory characteristics of the best treatment were evaluated using an incomplete block design with 15 semi-trained panelists. Attributes as colour, flavour, texture, appearance, and acceptability were considered. Three tastings sessions with a hedonic scale based on 5 points were developed.

2.9. Statistical analysis
Statistical analysis was performed with the GraphPad Prism 5.0 program (GraphPad Software, San Diego, California, USA) with a bidirectional analysis of variance. The test of comparisons was carried out with the Tukey test with a significance level of $P \leq 0.05$.

3. Results and discussion

3.1. Technological characteristics
Technological parameters such as height, slicing, and hardness are shown in table 2. In concordance to technological parameters, the greatest height value was observed in BQW501 (50% v/v whey and 10%

w/w of quinoa). Quinoa flour has a marked effect on height bread than wheat flour and was related with the percentages of quinoa incorporated to the formula when more quinoa was added the height was lower [16].

The sliced process showed that the best slicing sample was BQW503 treatment, which had 91.67%. The effect of quinoa in the hardness of the bread is highly significant. In contrast, the effect of whey was not significant. The interaction between quinoa and whey showed that increasing Quinoa flour produces an increase in the bread hardness (table 2).

Table 2. Technological parameters of bread.

Treatment	Height (cm)	Slicing (%)	Hardness (g)
BQW501	12.55±0.15[a]	73.33±0.58[a, b]	344.33±1.00[a]
BQW502	12.35±0.19[a]	76.67±1.00[b]	434.83±1.00[b]
BQW503	10.75±0.17[b]	91.67±0.58[c]	633.00±1.00[c]
BQW751	11.70±0.15[c]	80.00±1.15[d]	388.67±0.75[d]
BQW752	11.40±0.15[c,d]	71.67±0.58[a]	575.50±0.75[e]
BQW753	11.30±0.20[c, d]	75.00±0.58[b]	690.00±0.75[f]
BQW101	11.35±0.15[c, d]	91.67±1.15[c]	526.33±0.75[g]
BQW102	11.10±0.18[b, d]	83.33±1.15[e]	603.17±1.00[h]
BQW103	10.65±0.15[b]	73.33±1.15[a, b]	639.17±0.75[i]

Results are the mean ± standard deviation. One-way ANOVA: different letters (a,b) in the same column indicate significant differences between samples ($P \leq 0.05$).

3.2. Physicochemical properties

The analysis of the different pH values (table 3) showed that there is a significant influence of whey in the production of bread. However, such result was not found with the use of Quinoa. Despite the effect of whey on the pH, all treatments are within acceptable values according to standard NTE INEN 95 (INEN, 1979) which states that the pH of any bread should be between 5.5 and 6.

Total solids are significantly affected not only by the amount of quinoa but also the amount of whey when they are added to each treatment. The greater influence factor is whey since it was added in higher amounts (50% to 100%, w/w) compared to the quantity of quinoa (10% to 20% w/w). The moisture content in the different treatments was within the values established by the INEN (<35%).

Table 3. Physicochemical properties of bread.

Treatments	pH	Total solids %	Moisture %
BQW501	5.81±0.03[a]	81.37±0.56[a]	18.63±0.03[a]
BQW502	5.86±0.03[a]	83.27±0.57[b]	16.73±0.13[b, c]
BQW503	5.86±0.05[a]	83.58±0.13[b]	16.42±0.13[b,c]
BQW751	5.85±0.10[a]	82.24±0.12[a, b]	17.76±0.68[b]
BQW752	5.93±0.09[a]	86.05±0.68[c, d]	13.95±0.13[b]
BQW753	5.94±0.03[a]	87.61±0.82[c]	12.39±0.13[b]
BQW101	5.93±0.03[a]	85.22±0.74[d]	14.78±0.03[b]
BQW102	5.91±0.03[a]	86.78±0.13[c, d]	13.22±0.13[b]
BQW103	5.92±0.03[a]	86.60±0.56[c, d]	17.03±0.13[c]

Results are the mean ± standard deviation. One-way ANOVA: different letters (a,b) in the same column indicate significant differences between samples ($P \leq 0.05$).

3.3. Colour

The colour is one of the most important factors to determine consumer acceptance when working with partial or total substitutions of wheat flour. The effects of quinoa flour and whey on bread colour are shown in table 4. The colour parameters of a commercial bread were measured, with luminosity values (L) = +54; Axis green - red (a) = +16; Axis blue - yellow (b) = +35. L, a, and b values of quinoa bread were statistically different (p<0.05), and the amount of quinoa flour created a noticeable difference in all color parameters. The treatments were found in the circle of the first quadrant of the a + and b + scale, which allows establishing that the different treatments are between brown and yellow earth colours. The yellow colour of bread could be enhanced to lactose levels, which also resulted in Maillard reactions at a higher degree. Likewise, the amount of quinoa flour created a noticeable difference in the colour of bread, especially, the yellowness [17].

Table 4. Colour of bread.

Samples	**L**	**a**	**b**	**C**
BQW501	47.1 ± 0.3^{a}	$17.3 \pm 0.5^{a,b,c,d,e}$	$32.2 \pm 0.5^{a,c,f}$	$36.5 \pm 0.3^{a,b,c,d}$
BQW502	50.8 ± 0.8^{b}	$17.2 \pm 0.5^{a,b,c,d,e}$	$32.8 \pm 0.5^{a,f}$	$37.0 \pm 0.5^{b,c,d}$
BQW503	49.2 ± 0.3^{c}	$17.6 \pm 0.1^{b,c,d,e}$	$32.9 \pm 0.5^{b,d,f}$	$36.8 \pm 0.2^{a,b,c,d}$
BQW751	52.3 ± 0.7^{d}	16.7 ± 0.5^{c}	$34.7 \pm 0.5^{a,b}$	38.4 ± 0.5^{c}
BQW752	$48.0 \pm 0.8^{a,c}$	$17.9 \pm 0.6^{a,b,c,d}$	32.7 ± 0.6^{c}	$37.0 \pm 0,4^{a,b}$
BQW753	$50.8 \pm 0.2^{b,e}$	$17.9 \pm 0.2^{a,b,c,d}$	$34.6 \pm 0.5^{a,d,e}$	38.9 ± 0.3^{c}
BQW101	48.8 ± 0.2^{c}	$17.8 \pm 0.4^{a,b,c}$	$33.4 \pm 0.3^{d,e}$	37.8 ± 0.3^{d}
BQW102	50.9 ± 0.2^{e}	18.2 ± 0.1^{d}	$35.7 \pm 0.1^{b,f}$	39.5 ± 0.2^{c}
BQW103	53.2 ± 0.1^{f}	$16.7 \pm 0.6^{a,b,c,d}$	$36.2 \pm 0.2^{d,f}$	39.8 ± 0.2^{c}
Control	54.7 ± 0.1^{g}	16.1 ± 0.2^{e}	$35.2 \pm 0.1^{b,c,f}$	38.7 ± 0.3^{c}

[a-g] Different letters in the same column indicate significant differences between means (p < 0.05).

3.4. Proximal analysis

Nutritional characteristics of bread are shown in table 5. No significant differences were observed in the ash content (p>0.05). On the other hand, protein, humidity and lipids had significantly differences (p < 0.05). The control sample which includes only wheat flour and water, had the highest moisture content, while the sample BQW101 had the lowest value. Bread with quinoa flour had lower moisture content than other type of breads [18]. These differences could be attributed to the variation in soluble dietary fiber contents of quinoa flour. Presence of whey solids such as proteins (α-lactalbumine and β-lactoglobuline) and lipids, could explain higher protein and lipid values found in the sample BQW101 compared to the control. The replacement of water with pasteurized whey could have also increased the macro mineral concentrations and lipids of the bread [19].

Table 5. Nutritional properties of bread.

Sample	**Ash (%)**	**Protein (%) (Nx6.25)**	**Humidity (%)**	**Lipids (%)**
BQW101	2.40 ± 0.04^{a}	10.7 ± 0.7^{a}	29.9 ± 0.2^{a}	4.61 ± 0.1^{a}
Control	2.44 ± 0.04^{a}	9.94 ± 0.7^{b}	31.6 ± 0.2^{b}	3.30 ± 0.1^{b}

Results are the mean ± standard deviation. One-way ANOVA: different letters (a,b) in the same column indicate significant differences between samples (P ≤ 0.05).

3.5. Sensory evaluation

From the sensory evaluation of bread, it was possible to establish that at a 95% level of confidence there are significant differences between treatments. The sensory profile of treatments for the five characteristics showed that the highest acceptability was obtained by the treatment BQW101 (figure 1) which had a significant amount of whey and the highest amount of quinoa. On the other hand, the control and treatment BQW101 had the highest evaluations in texture. It can be noted that this is not affecting the acceptability of treatments. However, appearance is significantly affected by quinoa and whey.

Figure 1. Sensorial profile of bread.

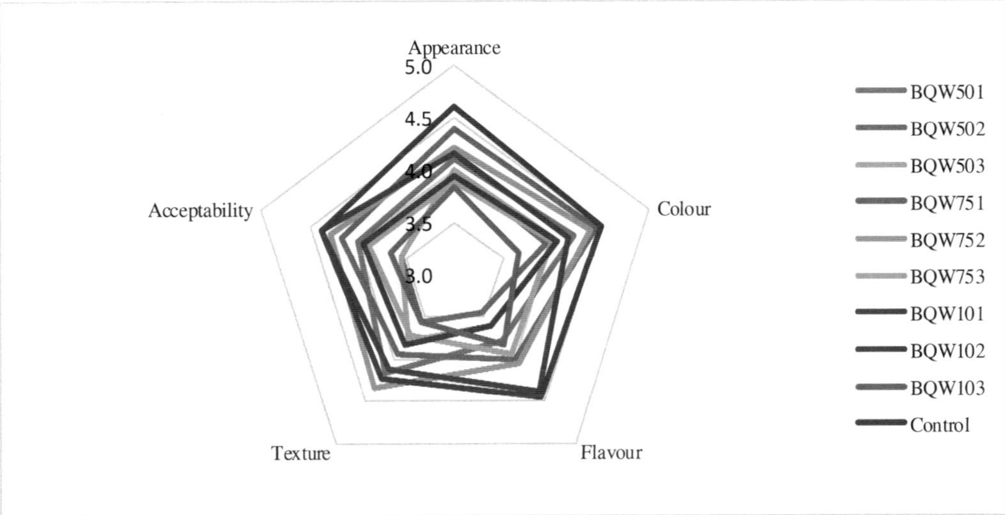

3.6. Microbiological analysis

The microbiological analysis establishes that the processed bread complies with the provisions of the INEN 2945; all the results are within the normative what indicate the absence of harmful microorganisms to public health. The shell life of the best treatment was obtained by linearization of the concentration of total aerobic organisms (ln (ufc/g) = 0.5685* time + 5.973; R^2 = 0.9649). The shelf life was calculated by the equation and was 5.69 days, which is an excellent value for bread with no preservatives added in its formulation.

4. Conclusion

The effect of quinoa flour and whey in bread showed that the maximum value of quinoa flour that can be added is 15%. Higher percentages cause loss of acceptability due to a slight bitterness of the bread. However, it helps the process of leavened as the sugar content in the flour provides sugars to the yeasts. In contrast, whey could completely replace the amount of water that is used to make bread. Whey does not affect the pH, height, texture, and generate greater acceptability in the sensory analysis because it hides unwanted flavours that could be attributed to the use of quinoa.

Acknowledgements
Financial support provided by Technical University of Ambato to participate in The 3rd International Conference on Agricultural and Biological Sciences (ABS2017) in Qingdao, Shandong Province, China.

References

[1] Hager A S, Zannini E and Arendt E K 2012 28 - Formulating breads for specific dietary requirements *Breadmaking* (2nd edi.) ed S P Cauvain (Cambridge, UK: Woodhead Publishing) pp 711-35

[2] Nowak V, Du J and Charrondière U R 2016 Assessment of the nutritional composition of quinoa (Chenopodium quinoa Willd) *Food Chem.* **193** 47-54
[3] Repo-Carrasco-Valencia R A-M and Serna L A 2011 Quinoa (Chenopodium quinoa, Willd) as a source of dietary fiber and other functional components *Food Sci. Tech-Brasil.* **31** 225-30
[4] Dewettinck K, Van Bockstaele F, Kühne B, Van de Walle D, Courtens T M and Gellynck X 2008 Nutritional value of bread: Influence of processing, food interaction and consumer perception *J. Cereal Sci.* **48(2)** 243-57
[5] Świeca M, Sęczyk Ł, Gawlik-Dziki U and Dziki D 2014 Bread enriched with quinoa leaves – The influence of protein–phenolics interactions on the nutritional and antioxidant quality *Food Chem.* **162** 54-62
[6] Giménez A, Varela P, Salvador A, Ares G, Fiszman S and Garitta L 2007 Shelf life estimation of brown pan bread: A consumer approach *Food Qual. Prefer.* **18(2)** 196-204
[7] Alvarez-Jubete L, Arendt E K and Gallagher E 2010 Nutritive value of pseudocereals and their increasing use as functional gluten-free ingredients *Trends Food Sci. Tech.* **21(2)** 106-13
[8] Ando H, Chen Y-C, Tang H, Shimizu M, Watanabe K and Mitsunaga T 2002 Food components in fractions of quinoa seed *Food Sci. Technol. Res.* **8(1)** 80-4
[9] Hirose Y, Fujita T, Ishii T and Ueno N 2010 Antioxidative properties and flavonoid composition of Chenopodium quinoa seeds cultivated in Japan *Food Chem.* **119(4)** 1300-6
[10] Schoenlechner R, Siebenhandl S and Berghofer E 2008 *7 - Pseudocereals. Gluten-Free Cereal Products and Beverages* (San Diego: Academic Press) pp 149-VI
[11] Wronkowska M, Jadacka M, Soral-Śmietana M, Zander L, Dajnowiec F, Banaszczyk P, Jelinski T and Szmatowicz B 2015 ACID whey concentrated by ultrafiltration a tool for modeling bread properties *LWT- Food Sci. Tech.* **61(1)** 172-6
[12] Kenny S, Wehrle K, Stanton C and Arendt E K 2000 Incorporation of dairy ingredients into wheat bread: Effects on dough rheology and bread quality *Eur. Food Res. Technol.* 391- 6
[13] Blaschek K M, Wendorff W L and Rankin S A 2007 Survey of salty and sweet whey composition from various cheese plants in Wisconsin *J. Dairy Sci.* **90(4)** 2029-34
[14] Dec B and Chojnowski W 2006 Characteristics of acid whey powder partially demineralised by nanofiltration *Polish J. Food Nutr. Res.* **56(SI 1)** 87-90
[15] Dissanayake M and Vasiljevic T 2009 Functional properties of whey proteins affected by heat treatment and hydrodynamic high-pressure shearing *J. Dairy Sci.* **92(4)** 1387-97
[16] De la Cruz W and Hugo, W 2009 Quinua (*Chenopodium quinoa* wild) and whey protein supplementation in wheat flour (*Triticum aestivum* L.) to extend the shelf life of mold bread. (Doctoral dissertation) National Agrarian University - La Molina (UNALM) Lima – Perú.
[17] Divya N and Rao K J 2010 Studies on utilization of Indian cottage cheese whey in wheat bread manufacture *J. Food Process Pres.* **34(6)** 975-92
[18] Alvarez-Jubete L, Arendt E K and Gallagher E 2010 Nutritive value of pseudocereals and their increasing use as functional gluten-free ingredients *Trends Food Sci. Tech.* **21(2)** 106-13
[19] Bilgin B, Daglioglu O and Konyali M 2006 Functionality of bread made with pasteurized whey and/or buttermilk *Ital. J. Food Sci.* **18(3)** 277-86

Response surface method optimization of ectoine fermentation medium with moderate halophilic bacteria *Halomonas* sp. H02

T T Li, A Qu, X N Yuan, F X Tan, X W Li, T Wang and L H Zhang[1]

Collaborative Innovation Center for Vessel Pollution Monitoring and Control @ Dalian Maritime University, Dalian 116026, China

E-mail: dlzlh2008@163.com

Abstract. Moderate halophilic bacteria are of halophilic bacteria whose suitable growth of NaCl is 5-10%. When the moderate halophilic bacteria response to high osmotic stress, the intracellular will synthesize small organic molecule compatible solutes. Ectoine, which is the major synthetic osmotic compatible solutes for moderate halophilic bacteria, can help microbial enzymes, nucleic acids and the whole cell resist to hypertonic, high temperature, freezing and other inverse environment. In order to increase the Ectoine production of Moderate halophilic bacteria *Halomonas* sp. H02, the Ectoine fermentation medium component was optimized by Plackett-Burman (PB) and Response Surface Methodology (RSM) based on the principle of non-complete equilibrium. The results of PB experiments showed that the three main influencing factors of Moderate halophilic bacteria *Halomonas* sp. H02 synthesis Ectoine culture medium were $C_5H_8NNaO_4$ concentration, NaCl concentration and initial pH. According to the center point of the steepest climbing experiment, the central combination design experiment was used to show that the model is consistent with the actual situation. The optimum combination of three influencing factors were $C_5H_8NNaO_4$ 41 g/L, NaCl 87.2 g/L and initial pH 5.9, and the predicted amount of Ectoine was 1835.8 mg/L, increased by 41.6%.

1. Introduction

Ectoine (1, 4, 5, 6-tetrahydro-2-methyl-4-pyrimidinecarboxylic acid) is one of the most important compatible solutes prevalent in most moderate halophilic bacteria [1-4]. Ectoine can not only equilibrate the osmotic pressure of cells, but also provide a reversible assistance on enzymes, DNA and the whole cell in inverse environments [5]. Therefore, it causes extensive attention from scholars, and the optimization of Ectoine synthesis conditions, especially the effect of fermentation medium on Ectoine synthesis has become a research hotspot in recent years

The effect of different carbon sources on Ectoine synthesis is related to the synthesis pathway of Ectoine. When $C_5H_8NNaO_4$ is the sole carbon source, it can provide a rich amino group for the reaction of transforming oxaloacetic acid into aspartic acid and transforming aspartic acid-β-semialdehyde into L-2,4-diaminobutyric acid during the Ectoine synthesis pathway, making the Ectoine synthesis higher [6]. The osmotic pressure of the culture medium has an important effect on the microbial growth activity, and the osmotic pressure of the appropriate cell growth is generally adjusted by adding an inorganic salt in the culture medium. In the range of NaCl concentration suitable for Ectoine synthesis, the higher the concentration of NaCl, the higher the concentration threshold, the higher the concentration of Ectoine in the intracellular [7, 8]. Therefore, higher concentrations of NaCl have a positive effect on the synthesis of Ectoine [9]. While each micro-organisms have their optimum pH, that pH is too low or too high not only affects the growth of micro-organisms, but also affects the product synthesis. The Deamination process of synthesis of Ectoine with $C_5H_8NNaO_4$ as the sole carbon source in the medium is a process of pH increase, so adjusting the appropriate initial pH has a certain effect on the synthesis of Ectoine.

In this paper, we used the response surface optimization method to establish the continuous variable surface model, at the same time optimize and evaluate the factors affecting the yield and their interactions. Therefore, the optimal conditions of the multi-factor system could be determined quickly and effectively [10]. In this paper, the main influencing factors of Ectoine fermentation medium were screened by PB design. Then, the optimum Ectoine synthesis area was selected according to the steepest climbing experiment. Finally, the optimal Ectoine fermentation medium was designed by center combination design, and the Ectoine synthesis was predicted under optimal conditions.

2. Materials and methods

2.1. Strains

Sample of strain: The strain *Halomonas* sp. H02 was isolated and identified by this laboratory.

2.2. Medium

Enrichment medium (LB medium) (g/L): peptone 10, yeast powder 5, NaCl 60, pH 7.2, 121 °C, 20 min sterilization. The medium was autoclaved at 121 °C for 20 min.

Induction medium (g/L): $C_5H_8NNaO_4\cdot H_2O$ 40, $(NH_4)_2SO_4$ 8, KH_2PO_4 3, K_2HPO_4 9, $MgSO_4\cdot 7H_2O$ 0.4, $MnSO_4\cdot H_2O$ 0.01, $FePO_4\cdot 2H_2O$ 0.01, $CH_3COONa\cdot 3H_2O$ 0.2, $Na_3C_6H_5O_7\cdot 2H_2O$ 0.05, NaCl 60, pH 7.2, trace mineral solution(EDTA-2Na 63.7, $ZnSO_4$ 2.2, $CaCl_2$ 5.5, $MnCl_2\cdot 4H_2O$ 5.1, $FeSO_4\cdot 7H_2O$ 5, $Na_2Mo_4\cdot 2H_2O$ 1.1, $CuSO_4\cdot 5H_2O$ 1.6, $CoCl_2\cdot 6H_2O$ 1.6) 2 mL, which contains 30% of the filtered seawater (the salt concentration of seawater is calculated as 2.75%). The medium was autoclaved at 121 °C for 20 min. The trace mineral solution was sterilized by filtration (0.22 μm pore size, Millipore Express, USA)

2.3. Ectoine induction and release

2.3.1. Ectoine induction. The strain *Halomonas* sp. H02 was inoculated into LB medium, and cultured at 30 °C and 120 rpm for 24 h. 1% inoculation quantity access to 30 mL induction medium, then at 30 °C, 120 rpm shake flask to induce culture for 72 h.

2.3.2. Release of Ectoine. Separate cells from fermentation broth. Using Induction medium, fermentation broth at the exponential phase of growth (72 h) was centrifuged at 4 °C and 12,000 rpm for 15 min.

Preparation of Ectoine sample: Cells were collected by the method described above and add 1 mL of 80% ethanol (v/v) resuspension, standing overnight. The suspension was centrifuged again (12000 rpm, 4 °C, 15 min) and the supernatant was stored for HPLC analysis.

2.4. Determination of Ectoine concentration

For the preparation of Ectoine samples, see "2.3.2".

Ectoine was determined by high performance liquid chromatography (HPLC) method. The sample for determining ectoine was prepared by ethanol extraction method [11]. We used TSK-GEL reversed phase column (TOSOH Corporation, Japan) with 50 mmol/L KPi buffer (pH 6) as the mobile phase at 35 °C. The flow rate of 1 mL/min and a UV detector wavelength of 210 nm was used [12]. The retention time of Ectoine was obtained by testing the Ectoine standard (the extracellular sample was water-soluble sample, and the intracellular sample was alcohol-soluble samples). Ectoine standards came from Biomol GmbH company (Germany).

Ectoine content calculation:

$$X = 2 \times \frac{S}{S_0} \times 142.2 \quad (1)$$

Where X represents Ectoine content, unit (mg/L), S represents Ectoine sample peak area and S_0 represents Ectoine standard peak area.

The standard concentration of Ectoine is 2 mol/L and the molecular weight of Ectoine is 142.2 g/mol.

2.5. 2PB experiment

According to the related literature and the results of the previous study of the laboratory, this experiment used N=12 PB design, selected seven factors (g/L), each factor selected high (+1) low (-1)two Level: X_1 $C_5H_8NNaO_4 \cdot H_2O$ (20, 40), X_2 $(NH_4)_2SO_4$ (6, 12), X_3 $FePO_4 \cdot 2H_2O$ (0.4, 0.8), X_4 $CH_3COONa \cdot 3H_2O$ (0.06, 0.12), X_5 $Na_3C_6H_5O_7 \cdot 2H_2O$ (0.3, 0.6), X_6 NaCl (75, 150), X_7 initial pH (6, 8), for other factors, see "2.2" induction medium. Experiment design was performed with Design-Expert.8.0 software. According to the results obtained, the effect of each factor on Ectoine level was analyzed and the significance was evaluated, then the factors which had a significant impact on Ectoine level were obtained.

2.6. The steepest climbing experiment

The response surface fitting equation was similar to the real situation in the immediate neighborhood of the study, therefore, it was necessary to approximate the maximum Ectoine production region to establish an effective response surface fitting equation [13]. The direction of the steepest climbing experiment and the step of change were determined by the three main influencing factors of $C_5H_8NNaO_4$ concentration, NaCl concentration and initial pH according to the results of PB experiment, and the area of Ectoine maximum production could be approached rapidly.

2.7. Center combination experiment

According to the factors determined by the PB experiment and the center point of the steepest climbing experiment, three factors (X_1-sodium chloride ($C_5H_8NNaO_4$), X_2-sodium acetate (NaCl), X_3-initial pH) were designed by using the Central composite design (CCD) in Design-Expert.8.0 software. The experiment took the level of Ectoine as a response value, while taking five levels, encoded as (-1.68, -1, 0, 1, 1.68). The experiment design was shown in table 1.

Table 1. Level and code of variables chosen for Central composite rotatable design.

Variable	Coded level				
	-1.68	-1	0	1	1.68
X_1(g/L)	23.18	30.00	40.00	50.00	56.82
X_2(g/L)	64.77	75.00	90.00	105.00	115.23
X_3	5.16	5.50	6.00	6.50	6.84

3. Result and analysis

3.1. Ectoine synthesis of key influencing factors

According to the method "2.5" design PB experiment, PB experiment results were shown in table 2. The variance analysis of table 2 experiment data was carried out by Design-Expert.8.0 software. The results were shown in table 3.

IOP Conf. Series: Earth and Environmental Science **77** (2017) 012019 doi:10.1088/1755-1315/77/1/012019

Table 2. Plackett-Burman experiment design and results.

Serial number	X_1	X_2	X_3	X_4	X_5	X_6	X_7	X_8	X_9	X_{10}	X_{11}	Ectoine (mg/L)
1	1	1	1	-1	-1	-1	1	-1	1	1	-1	976.5
2	-1	1	1	-1	1	1	1	-1	-1	-1	1	426.6
3	-1	-1	-1	1	-1	1	1	-1	1	1	1	417.0
4	-1	-1	-1	-1	-1	-1	-1	-1	-1	-1	-1	830.8
5	1	-1	1	1	1	-1	-1	-1	1	-1	1	1268.7
6	1	-1	-1	-1	1	-1	1	1	-1	1	1	1028.4
7	-1	-1	1	-1	1	1	-1	-1	-1	-1	1	593.1
8	1	1	-1	1	1	1	-1	-1	-1	1	-1	785.3
9	1	-1	1	1	-1	1	1	1	-1	-1	-1	641.5
10	1	1	-1	-1	-1	1	-1	1	1	-1	1	984.2
11	-1	1	-1	1	1	-1	1	1	1	-1	-1	649.8
12	-1	1	1	1	-1	-1	-1	1	-1	1	1	752.4

Table 3. Analysis of variance of PB results.

Source	Value	F	Prob>F
X_1	167.92	40.07	0.0032
X_2	-17.06	0.41	0.5551
X_3	-3.06	0.013	0.9138
X_4	-27.08	1.04	0.3651
X_5	12.48	0.22	0.6626
X_6	-138.24	27.15	0.0065
X_7	-89.57	11.40	0.0279

It can be seen from table 3 that according to the variance analysis results of PB design, the main influencing factors of Ectoine synthesis were determined as $C_5H_8NNaO_4$ concentration, NaCl concentration and initial pH of the culture medium. Which $C_5H_8NNaO_4$, NaCl on the confidence of Ectoine synthesis in more than 95% had a significant impact. The initial pH of the culture medium had a significant effect on the confidence of Ectoine synthesis in more than 90%. Meanwhile, according to table 3, in addition to $C_5H_8NNaO_4$ and $Na_3C_6H_5O_7$ showed a positive effect, other factors showed a negative effect on Ectoine synthesis. The next step in the optimization should increase the concentration of $C_5H_8NNaO_4$ appropriated in the fermentation medium, reduced the concentration of NaCl and initial pH, in addition to $Na_3C_6H_5O_7$, other factors selected the low level of concentration.

3.2. The steepest climbing experiment

In the steepest climbing experiment, the change range of the concentration of $C_5H_8NNaO_4$, NaCl and initial pH were selected as 10 g/L, 15 g/L and 0.5, and the combination of the maximum concentration of Ectoine was selected. The design and results of the steepest climbing experiment were shown in table 4.

The results showed that the Ectoine level reached 1679.2 mg/L under No. 3 experiment, and the level of Ectoine decreased with the concentration. Therefore, No. 3 experiment was selected as the central point of the central combination experiment.

Table 4. The design and results of the steepest climbing test.

Serial number	X_1 (g/L)	X_6 (g/L)	X_7	Ectoine (mg/L)
1	20	60	5.5	867.8
2	30	75	6	1114.7
3	40	90	6.5	1679.2
4	50	105	7	1322.8
5	60	120	7.5	931.8

3.3. Response surface analysis of Ectoine fermentation medium components

3.3.1. Quadratic regression model fitting and verification.

The $C_5H_8NNaO_4$ Concentration, NaCl Concentration and initial pH of the culture medium were determined in 3.2, and the amount of Ectoine synthesis was used as the response value. The experiment results were shown in tables 5 and 6.

Table 5. Design and results of the center combination experiment.

Serial number	X_1	X_2	X_3	Ectoine (mg/L)
1	30.00	75.00	5.50	1300.3
2	40.00	90.00	6.00	1887.8
3	30.00	105.00	5.50	912.8
4	56.82	90.00	6.00	1341.6
5	30.00	105.00	6.50	1240.9
6	50.00	105.00	6.50	947.6
7	40.00	90.00	6.00	1900.0
8	50.00	105.00	5.50	1132.8
9	40.00	64.77	6.00	1261.6
10	40.00	90.00	6.00	1849.0
11	40.00	115.23	6.00	995.4
12	40.00	90.00	6.00	1870.0
13	50.00	75.00	5.50	1457.7
14	40.00	90.00	5.16	1583.3
15	40.00	90.00	6.84	1387.0
16	30.00	75.00	6.50	1095.6
17	40.00	90.00	6.00	1876.6
18	23.18	90.00	6.00	1159.9
19	40.00	90.00	6.00	1912.9
20	50.00	75.00	6.50	986.2

Table 6. Analysis of variance of the results of the central combination experiment.

Source	df	Mean Square	*F*	*Prob>F*
Model	9	2.698E+005	41.13	<0.0001
X_1	1	5748.95	0.88	0.3713
X_6	1	81279.71	12.39	0.0055
X_7	1	54599.79	8.32	0.0162
$X_1 X_6$	1	1841.03	0.28	0.6078
$X_1 X_7$	1	76081.20	11.60	0.0067
$X_6 X_7$	1	83836.94	12.78	0.0051
X_1^2	1	8.885E+005	135.44	<0.0001
X_6^2	1	1.225E+006	186.68	<0.0001
X_7^2	1	3.944E+005	60.11	<0.0001
Residual	10	6560.35		
Lack of fit	5	12604.75	24.43	0.0016
Pure error	5	515.96		
Cor total	19	2.494E+006		
R^2=0.9737				

The experiment results regression analysis was performed by Design-Expert.8.0 software to fit out a quadratic equation: Y(mg/L)=1886.52 + 20.52 X_1 – 77.15 X_2 – 63.23 X_3 – 15.17 X_1X_2 – 97.52 X_1X_3 + 102.37 X_2X_3 – 248.31 X_1^2 – 291.52 X_2^2 – 165.42 X_3^2.

The variance analysis of the model was shown as table 6. The confidence interval of the model was within 99%, the confidence is 99.99%, and the regression equation is significant, indicating that the equation has a good fit degree; the equation is consistent with the actual situation. In this model, the confidence levels of X_2, X_3, X_1X_3, X_2X_3, X_1^2, X_2^2, X_3^2 coefficients were all above 95%, indicating that they were all significant items of this model. The coefficient of determination of the model was R^2 = 0.9737, indicating that the model was in good agreement with the actual situation.

3.3.2. Response surface analysis and determination of the best medium components.

Combined with the regression equation, the response surface analysis diagram was plotted by the Design-Expert.8.0 software. Each response surface plot represented the effect of two independent variables, holding the other variables at zero [14].

From the three-dimensional response surface (figures 1-3) showed that there was a real maximum for the fitting surface, that all the three factors had a more appropriate concentration. In order to further determine the stability of the Ectoine maximum production, the first order partial derivative of the regression equation is made equal to zero. The results show as:

$$20.52-15.17X_2-95.72X_3-496.62X_1=0 \quad (2)$$

$$-71.15-15.17X_1+102.37X_3-583.04X_2=0 \quad (3)$$

$$-63.23-95.72X_1+102.37X_2-330.84X_3=0 \quad (4)$$

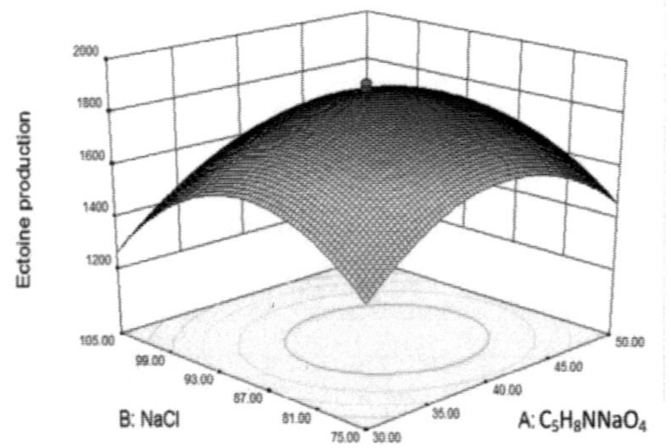

Figure 1. Response surface of $C_5H_8NNaO_4$ and NaCl on the Ectoine production.

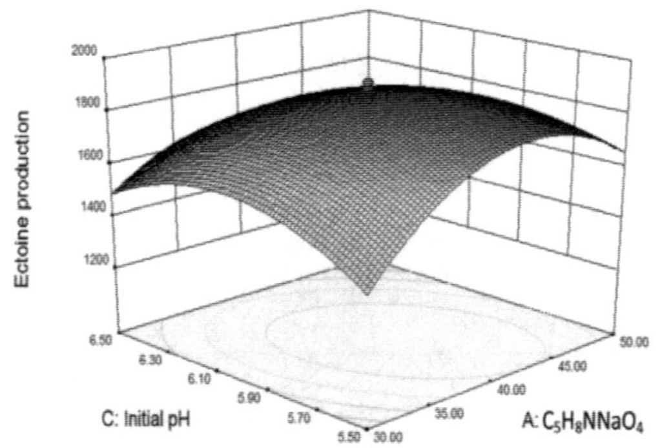

Figure 2. Response surface of $C_5H_8NNaO_4$ and initial pH on the Ectoine production.

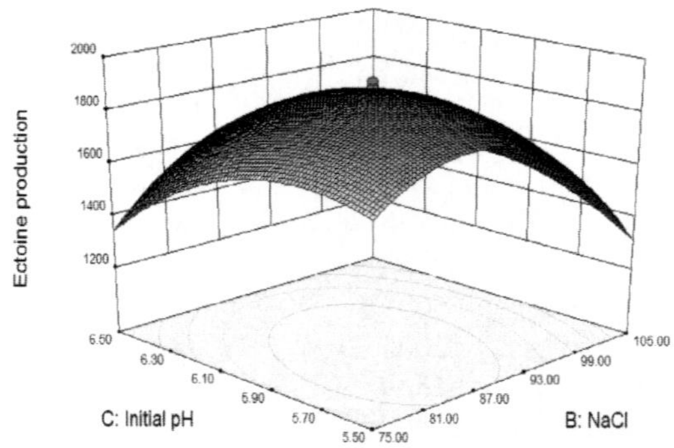

Figure 3. Response surface of NaCl and initial pH on the Ectoine production.

The extreme point of the model was obtained by solving the equation group: X_1=-0.102, X_2=0.184, X_3=0.278; Substituting the regression equation, the predicted Ectoine production was 1835.8 mg/L. The concentrations of $C_5H_8NNaO_4$ and NaCl were 41 g/L, 87.2 g/L, respectively, and the initial pH was 5.9.

In order to confirm the predicted results of the experiment, the fermentation experiments were carried out with the optimum medium conditions obtained by response surface experiment. Repeat the experiment twice, the concentrations of Ectoine were 1765.5 mg/L and 1862.4 mg/L, respectively, and the average value was 1813.9 mg/L. The difference between the experimental value and the predicted value is very small, indicating that the model can predict the actual fermentation situation appropriately.

4. Conclusion

This paper uses the Plackett-Burman design in the first place, with the amount of Ectoine synthesis as the response value, three main factors influencing the amount of Ectoine synthesis was screened out from seven factors, and they were $C_5H_8NNaO_4$ concentration, NaCl concentration and initial pH of the culture medium. Through the steepest climbing test, No. 3 experiment was chosen as the central point of the center combination test. The optimum combination of three influencing factors was determined as $C_5H_8NNaO_4$ 41 g/L, NaCl 87.2 g/L and initial pH 5.9 through the central combination experiment.

The optimal fermentation medium for Ectoine produced by *Halomonas* sp. H02 were (g/L): $C_5H_8NNaO_4 \cdot H_2O$ 41, $(NH_4)_2SO_4$ 8, KH_2PO_4 3, K_2HPO_4 9, $MgSO_4 \cdot 7H_2O$ 0.4, $MnSO_4 \cdot H_2O$ 0.01, $FePO_4 \cdot 2H_2O$ 0.4, $CH_3COONa \cdot 3H_2O$ 0.06, $Na_3C_6H_5O_7 \cdot 2H_2O$ 0.6, NaCl 87.2, initial pH 5.9, trace elements 2 mL. Which contains 30% of the filtered seawater (sea salt concentration by 2.75%). This medium was named as ZMG medium.

The verification experiment to this model was carried out on the optimized medium, and the amount of Ectoine synthesis reached 1835.8 mg/L, which was 41.6% higher than before optimization, indicating that it is scientifically reasonable and accurate to combine Plackett-Burman experiment and response surface in optimizing the Ectoine fermentation medium of *Halomonas* sp. H02.

Acknowledgments

This research was supported by Higher Education Development Fund (for Collaborative Innovation Center) of Liaoning Province, China [20110216001] and the Collaborative Innovation Center for Vessel Pollution Monitoring and Control @ Dalian Maritime University.

References

[1] Ventosa A, Nieto J J and Oren A 1998 Biology of moderately halophilic aerobic bacteria *Microbiol. Mol. Biol. Rev.* **62(2)** 504-44

[2] Oren A 2002 Diversity of halophilic microorganisms: Environments, phylogeny, physiology, and applications *J. Ind. Microbiol. Biot.* **28** 56-63

[3] Ren P G and Zhou P J 2003 Advances in studies on moderate halophilic bacteria *Acta Microbiol. Sin.* **43(3)** 427-31

[4] Wang Y O and Tao T 1994 Compatible solute in microbial osmotic pressure regulation *Microbiol. Chin.* **21(5)** 293-6

[5] Zheng X, Ma H and Yan X W 2010 Ectoine synthesis and release under osmotic shock in halomonas venusta DSM4743 *Microbiol. Chin.* **37(7)** 1090-6

[6] Zhang L H and Lang Y J 2008 Method for producing Ectoine by microbial fermentation (China: CN101314785)

[7] Grammann K, Volke A and Kunte H J 2002 New type of osmoregulated solute transporter identified in halophilic members of the Bacteria domain: TRAP-transporter TeaABC mediates uptake of ectoine and hydroxyectoine in Halomonas elongata DSM 2581T *J. Bacteriol.* **184** 3078-85

[8] Kuhlmann A U and Bremer E 2002 Osmotically regulated synthesis of compatible solute ectoine in Bacillus pasteurii and related Bacillus spp *Appl. Environ. Microb.* **68(2)** 772-83

[9] Tao P, Li H, Yu Y, Gu J and Liu Y 2016 Ectoine and 5-hydroxyectoine accumulation in the halophile Virgibacillus halodenitrificans PDB-F2 in response to salt stress *Appl. Environ.*

Microb. **100(15)** 6779-89

[10] Van-Thuoc D, Guzmán H, Mai T H and Hatti-Kaul R 2010 Ectoine production by Halomonas boliviensis: Optimization using response surface methodology *Mar. Biotechnol.* **12(5)** 586-93

[11] Zhang L H, Lang Y J and Nagata S 2009 Efficient production of ectoine using ectoine-excreting strain *Extremophiles* **13** 717-24

[12] Lang Y J, Bai L, Ren Y N, Zhang L H and Nagata S 2011 Production of ectoine through a combined process that uses both growing and resting cells of Halomonas salina DSM 5928T *Extremophiles* **15(2)** 303-10

[13] Khuri A I and Mukhopadhyay S 2011 *Response Surface Methodology* (Berlin Heidelberg: Springer)

[14] Zhou J, Yu X, Ding C, Wang Z, Zhou Q, Pao H and Cai W 2011 Optimization of phenol degradation by Candida tropicalis Z-04 using Plackett-Burman design and response surface methodology *J. Environ. Sci-China* **23(1)** 31-6

IOP Conf. Series: Earth and Environmental Science **77** (2017) 012020 doi:10.1088/1755-1315/77/1/012020

The characteristics of phosphorus adsorption and desorption in gray desert soil of Xinjiang, China

B Wang[1], J S Sun[1, 3], H Liu[1] and Y B Ma[2]

[1] Institute of Soil Fertilizer and Agricultural Water Saving, Xinjiang Academy of Agricultural Sciences, Urumqi 830091, China
[2] National Soil Fertility and Fertilizer Effects Long–term Monitoring Network, Institute of Agricultural Resources and Regional Planning, Chinese Academy of Agricultural Sciences, Beijing 100081, China

E-mail: 550904041@qq.com

Abstract. The characteristics of phosphorus (P) adsorption and desorption in Xinjiang gray desert soil (0 - 200 mm) of China in the long-term fertilization condition is affected by the level of soil P content which studied through an isothermal adsorption and desorption experiments of P. The results stated that within the experimental concentration range, with the increase of the amount of outer-source phosphorus, P adsorption, desorption and desorption rate increased and adsorption rate decreased gradually in different Olsen-P levels of gray desert soil in Xinjiang, China. Olsen-P content is significantly correlated with the P adsorption saturation (*DPS*) of gray desert soil. The maximum adsorption capacity (X_m) of the treatments followed an extremely significant decreasing order of CK>NPK≈NPKM>PK≈NPKS. The maximum buffer capacity (*MBC*) and adsorption constant (*K*) of the NPK treatment was much higher than NPKM, NPKS, PK and CK treatments. And, *MBC* value of CK treatment was extremely higher than NPKS and PK, however, the differences between NPKM and CK, NPKS and PK were not significant. The comparison between NPKM, NPKS, PK and CK treatments showed no significant difference in *K* value, but these four showed significantly lower than NPK treatments. The value of soil easy desorption P (*RDP*) of NPKS and NPKM was significantly higher than NPK and PK, and the chemical fertilizer with organic fertilizer was a best way to release the phosphorus for Xinjiang agricultural production, China.

1. Introduction

Phosphorus (P) is an indispensable nutrient for plant growth and is important for agricultural production [1-4]. The content of available P is relatively low in most soils in China, so meeting the nutrient requirements of plants is difficult. The calcareous soil in northern China has strong adsorption and precipitation effects on P fertilizer; thus, many studies have focused on methods to decrease fixation and increase the effectiveness of added P in soil [5]. Since the 1950s, efforts have been made to understand the adsorption and desorption characteristics of soil P from various aspects. Boparai and Sharma [6] demonstrated that clay type and its content, and content of available P in soil (Olsen-P), played important roles in the adsorption and desorption of soil P. Chinese researchers showed that soil pH, iron and aluminum oxides, calcium, organic matter and clay content all affected the P adsorption and desorption of soil [7]. The adsorption and desorption characteristics of P in soils with diverse particle compositions [5], different clay contents [4] and various soil types [8] were significantly

IOP Conf. Series: Earth and Environmental Science **77** (2017) 012020 doi:10.1088/1755-1315/77/1/012020

different. In addition, exogenous application of humic acid [9] and polyacrylamide [10] greatly influenced the P adsorption and desorption of soil. Even using the same soil type (such as gray desert soil) under similar climate conditions, the content of Olsen-P showed significant differences of up to 30 $mg \cdot kg^{-1}$ for different fertilization treatments in long-term experiments [11]. With such a great variation in the Olsen-P content of gray desert soil, what are the adsorption and desorption characteristics of soil P? What is the effect of different Olsen-P contents on the adsorption and desorption of soil P? Based on 22-year long-term fertilization experiments with gray desert soil, we studied the adsorption and desorption characteristics of soil P. This provided a basis for the appropriate utilization and management of P, as well as for optimizing fertilization in gray desert soil, Xinjiang China.

2. Materials and methods

2.1. Experimental field, climate characteristics and soil properties

Gray Desert Soil Long-term Monitoring Station in Urumqi Xinjiang is one station in the national long-term monitoring network that monitors soil fertility and fertilizer benefits. It was founded in 1989, and its primary goal is to investigate the climate characteristics, crop rotation, the effects of organic and inorganic fertilizer on crop yield, and the change in soil fertility in the northwestern agricultural region in China. The station (43°58′N, 87°25′E, with the altitude of 600 m) is located in a suburb of Urumqi, Xinjiang, in a typical continental arid climate zone. The climate characteristics of the area are as follows: annual precipitation, 242 mm; average temperature, 7.6°C; effective accumulated temperature, 1734°C; frost-free period, 156 d; and sunshine duration, 2454 h. The soil classification standard of China classes the soil in this station as typical gray desert soil; based on the standard of the United Nations Food Agriculture Organization (FAO) classification, it is a Calcaric Cambisol [12]. Before the fertilization experiments, the main soil properties in the plough layer (0-200 mm): soil pH (water: soil of 2.5:1) 8.1, organic matter 8.8 $g \cdot kg^{-1}$, total nitrogen (N) 0.87 $g \cdot kg^{-1}$, total P 0.67 $g \cdot kg^{-1}$, total potassium (K) 23.0 $g \cdot kg^{-1}$, alkaline hydrolysable N 55.2 $mg \cdot kg^{-1}$, Olsen-P 3.9 $mg \cdot kg^{-1}$, and NH_4OA_c extractable K 288.0 $mg \cdot kg^{-1}$. In 1989, prior to beginning the long-term fertilization experiments, a local cultivar of wheat was grown for 2 years (1988 and 1989) without fertilizer in order to reduce soil variation among treatments. Each following year, plants were grown for one season at the station with a rotation of wheat, maize and cotton.

In the present study, five treatments with large differences in Olsen-P were selected from the long-term fertilization experiment as shown in table 1 to study the adsorption and desorption of soil P. The five treatments were (1) control (CK), (2) P and K fertilization (PK), (3) N, P and K fertilization (NPK), (4) N, P, K and organic fertilization (NPKM) and (5) N, P and K fertilization with straw returning (NPKS). Winter wheat was planted in 2011. Fertilization in each treatment is shown in table 2.

Table 1. Soil properties (0–200 mm) of different treatments in the experiment.

Treatments	pH	Organic matter ($g \cdot kg^{-1}$)	Total N ($g \cdot kg^{-1}$)	Total P ($g \cdot kg^{-1}$)	Available N ($mg \cdot kg^{-1}$)	Olsen-P ($mg \cdot kg^{-1}$)	Available K ($mg \cdot kg^{-1}$)	Bulk density ($g \cdot cm^{-3}$)
CK	9.16	12.70	0.54	0.58	38.82	4.27C	216.80	1.56
PK	9.62	12.48	0.48	0.80	40.09	22.39B	222.04	1.50
NPK	9.26	13.30	0.58	0.92	70.64	13.61BC	284.90	1.50
NPKM	9.26	14.96	0.81	0.90	85.92	45.42A	939.96	1.46
NPKS	9.26	13.41	0.69	0.72	51.55	8.93BC	291.97	1.52

Note: The topsoil sample collected in June 2011 was used for table 1. The indicators are the soil fertility properties determined prior to the adsorption and desorption experiments. Different capital letters following numbers represent significant differences at $P < 0.01$, the same as below.

Table 2. Average annual nutrient input from different sources during 1990–2011.

Treatment	Chemical fertilizer			Manure	Straw
	N (kg·ha^{-1})	P (kg·ha^{-1})	K (kg·ha^{-1})	M (t·ha^{-1})	S (t·ha^{-1})
CK	0	0	0	0	0
PK	0	29.2–60.2	18.8–50.4	0	0
NPK	99.4–241.5	29.2–60.2	18.8–50.4	0	0
NPKM	29.8–84.9	8.7–22.4	6.7–10.1	30	0
NPKS	89.4–216.7	24.5–50.8	16.9–42.3	0	4.5–9

Note: Crops were planted without fertilizing in CK; the amount of P and K fertilizer in PK treatment was the same as that in NPK treatment; the fertilization amounts of N, P and K were 99.4, 29.2 and 18.8 kg·ha^{-1}, respectively, for 1990–1994, and were increased to 241.5, 60.2 and 50.4 kg·ha^{-1} for 1994–2011. In Treatment NPKM, the organic manure was sheep droppings, of which the nutrient contribution was calculated on the basis of 15%, the utilization rate of organic manure in the current year and in Treatment NPKS, the nutrient contribution of the straw was calculated on the basis of 10%, the utilization rate of straw in the current year. The two treatments were designed to have the same nutrient application rates as the same as treatment NPK. Of the chemical fertilizers, all the P and K fertilizers and 60% of the N fertilizer was used as base fertilizer, and 40% of the N fertilizer as topdressing.

2.2. Evaluation items

2.2.1. Determination of soil P with isothermal adsorption. Six soil samples (2.500 g each) were placed in 100-mL centrifuge tubes, and 25 mL of 0.01 mol·L^{-1} $CaCl_2$ solution with P concentrations of 0, 25, 50, 75 and 100 mg·L^{-1} respectively added into the six tubes. In addition, three drops of chloroform were added to inhibit microbial activity. The mixture was oscillated for 1 h at 25°C and centrifuged at RCF 2770 (×g) for 8 min after incubating for 24 h and then the P concentration of the solution was determined, based on which the P adsorption in soil was calculated. The curve of isothermal adsorption was plotted using the P concentration in the solution as the X-axis and adsorptive P in soil as the Y-axis, and then fitted with an equation.

2.2.2. Determination of soil P with isothermal desorption. The supernatant of the centrifuged from the mixture was removed, and the pellet rinsed twice with 25 mL of saturated NaCl solution and centrifuged at RCF 2770 (×g) for 8 min to eliminate free P. In each tube, 25 mL of 0.01 mol·L^{-1} $CaCl_2$ solution was added, with three drops of chloroform to inhibit microbial activity. The mixture was incubated for 24 h after oscillating for 1 h at 25°C, followed by centrifugation at RCF 2770 (×g) for 8 min. The resulting supernatant was used to evaluate the desorptive P, and the desorption rate was determined according to the ratio of desorptive P and total adsorptive P in the soil.

2.2.3. Determination of available P and organic matter in soil. Olsen-P in soil was determined using the 0.5 mol·L^{-1} $NaHCO_3$ (pH 8.5) method. The organic matter in soil and other items were evaluated using the conventional methods of soil agrochemistry [13].

2.3. Data processing

The maximum buffering capacity of soil P (MBC, mL·g^{-1}) was calculated using constant of adsorption reaction (K) × X_m [14], where X_m is the maximum adsorption of soil P. Readily desorbable P (RDP) represented the P amount from the soil solid phase entering the liquid phase with 0 mg·L^{-1} P addition during the adsorption and desorption process [14]. Statistical analysis was performed using SAS 9.1 and Excel 2003.

3. Results and analyses

3.1. Adsorptive characteristics of soil P

3.1.1. Isothermal adsorption curve of soil P. The adsorption curve in five treatments with various levels of soil P is shown in figures 1 and 2. With P addition in the range of 0–100 mg·L^{-1}, the P adsorption was gradually elevated and the absorptive rate was reduced with increases in exogenous P. With P addition in the range 0–50 mg·L^{-1}, the adsorption rapidly increased in all treatments except for PK; however, within the range 50–100 mg·L^{-1}, P adsorption in soil decreased in all five treatments. In all treatments with different P levels, the adsorption rapidly increased, then slowly increased, followed by a plateau with increased exogenous application of P, but the adsorption rate showed a downward trend. To quantify the variation of P adsorption in different treatments, the isothermal adsorption equation was selected to simulate increasing exogenous P and P adsorption in soil, which facilitates study of the adsorptive characteristics of soil P under various P treatments.

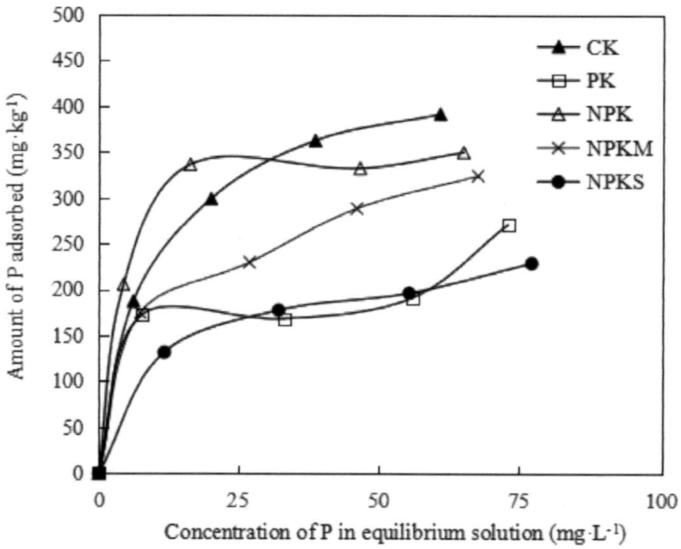

Figure 1. The P isotherm adsorption curves in different treatments.

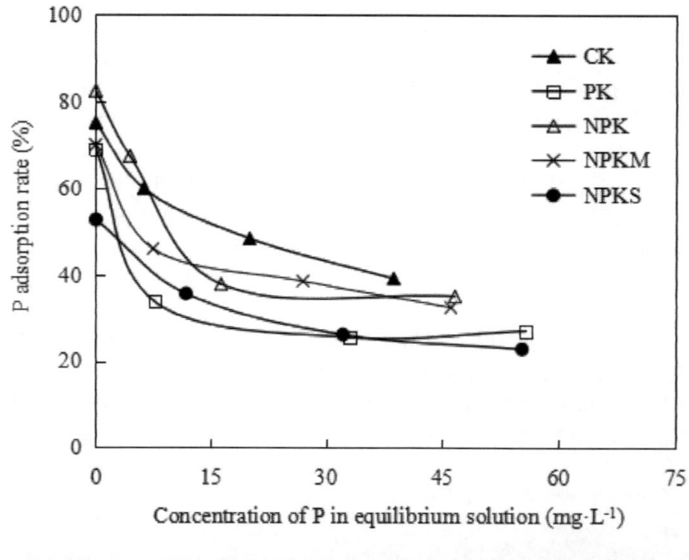

Figure 2. Change in P adsorption rate with extraneous P.

3.1.2. Adsorption equations of soil P. When using the equilibrium method to study P adsorption in soil systems, isothermal adsorption equations, such as Langmuir, Freundlich and Temkin [15], are commonly used to fit the relationship between the adsorption amount of the soil surface and the equilibrium concentration in solution. Then, parameters such as the adsorption amount can be calculated according to each equation. In the present study, we used the Langmuir, Freundlich and Temkin isothermal equations a shown in table 3 to simulate the isothermal adsorption process of soil P under various P levels.

The Langmuir equation is expressed as $C/X = C/X_m + 1/(KX_m)$, where X represents the P adsorption of soil, C indicates P concentration in the equilibrium solution (The parameter of X and C are the same definition for all three equations) and K is a constant related to the energy of adsorption. The Freundlich equation is $X = KC^b$, where K and b are two constants related to soil properties. The Temkin equation is $X = K_1 \ln C + K_1 \ln K_2$, where K_1 and K_2 are two constants related to soil properties. Simulation of soil P adsorption under the five treatments was carried out using the three isothermal adsorption equations as shown in table 3. The significance level of Langmuir equation fitting ($P < 0.01$ or $P < 0.05$) was higher than that of Freundlich and Temkin equations; and it showed good correlations, with all the correlation coefficients (r) > 0.95 as shown in table 4. This suggested that the Langmuir equation was the best method to describe the adsorption characteristics of P in gray desert soil, and so we selected this equation to explore the adsorption characteristics of P in soil.

Table 3. Isothermal adsorption equations and parameters in different treatments.

Treatment	Langmuir adsorption equation $C/X = C/X_m + 1/KX_m$			Freundlich adsorption equation $X = KC^b$			Temkin adsorption equation $X = K_1 \ln C + K_1 \ln K_2$		
	X_m	K	r	b	K	r	K_1	K_2	r
CK	416.7	0.2131	0.99**	0.3305	105.6	0.99**	91.20	0.2540	1.00**
PK	244.6	0.1807	0.95*	0.1441	119.5	0.65	30.72	3.0704	0.64
NPK	333.6	0.6086	0.99**	0.1572	177.5	0.79	41.24	4.0511	0.76
NPKM	330.2	0.1969	0.99**	0.2817	96.7	0.99*	67.01	0.4580	0.97*
NPKS	236.1	0.2445	0.99**	0.2834	65.4	0.99**	48.98	0.1766	0.99*

Note: * and ** represent statistical significance at $P < 0.05$ and $P < 0.01$, respectively. The same as below.

Table 4. Langmuir isothermal adsorption equations in different treatments.

Treatment	Langmuir equation	r
CK	$C/X = 0.0024C + 0.0113$	0.99**
PK	$C/X = 0.0041C + 0.0235$	0.95*
NPK	$C/X = 0.003C + 0.0049$	0.99**
NPKM	$C/X = 0.003C + 0.0178$	0.98**
NPKS	$C/X = 0.0043C + 0.0315$	0.98**

3.1.3. Adsorption parameters of soil P. The adsorption process of P can be characterized by many parameters. The following four parameters were used in the present study:

- X_m reflects the number of adsorption sites of soil colloids [16] and indicates the P storage capacity of soil. When the soil P storage reaches a certain capacity, it is possible for the soil to provide nutrients to crops [4]. Of the five treatments, X_m was the largest for CK, 12–18 times those of other treatments, and it was significantly different ($P < 0.01$) from other treatments as shown in table 5. There was no significant difference in X_m between NPK and NPKM, and between PK and NPKS; but for NPK and NPKM, there was a much greater X_m compared with that of PK and NPKS treatments. The P input of PK and NPK treatments were the same as

IOP Conf. Series: Earth and Environmental Science **77** (2017) 012020 doi:10.1088/1755-1315/77/1/012020

shown in table 2, but the yield of PK was substantially lower than that of NPK due to the lack of N [17] – that is, the output of soil P was less than that of the NPK treatment, resulting in a relatively high P left over in soil as shown in table 1. Therefore, the soil adsorption capacity was relatively small in the PK treatment. Similarly, the output of soil P was lower in the NPKS than in the NPK treatment, and so the soil adsorption capacity was relatively small.

Table 5. The P adsorption and desorption parameters in different treatments.

Treatment	Maximum adsorption capacity (X_m) ($mg \cdot kg^{-1}$)	Adsorption constant (K)	Max buffering capacity of soil P (MBC) ($mL \cdot g^{-1}$)	P sorption saturation (DPS) (%)	Readily desorbable P (RDP) ($mg \cdot kg^{-1}$)
CK	416.7A	0.21B	88.78B	1.02D	0.00C
PK	244.6C	0.18B	43.90C	8.83B	0.25C
NPK	333.6B	0.61A	202.77A	4.10C	0.00C
NPKM	330.2B	0.20B	64.37BC	14.24A	3.82A
NPKS	236.1C	0.24B	53.37C	3.87CD	1.39B

- The constant K represents the affinity between soil colloids and phosphate ions, which is a strength factor of soil adsorption [18], reflecting the level of P adsorption of soil to a certain extent. A positive K-value indicates that the adsorption reaction is spontaneous at room temperature, and the magnitude of K reflects the spontaneous extent of the adsorption reaction. The larger the K-value, the stronger the spontaneity and the more stable the products are, leading to weak P supply [4]. The K-values were positive in all treatments, indicating that the adsorption reactions were all spontaneous in the order of NPK > NPKS > CK > NPKM > PK. Of these, the K-value was significantly higher for the NPK treatment compared with the other treatments, and no substantial difference was observed among the other four treatments, suggesting that the soil P adsorption was increased after applying N, P, and K fertilizer, but the application of P and K, NPK with organic fertilizer or straw reduced the P adsorption of soil.
- MBC is the product of K and X_m in the Langmuir equation (i.e. $MBC = KX_m$) and so is a combined parameter for the two factors of intensity of P adsorption (K) and capacity (X_m), and it is a *comprehensive* indicator to evaluate the soil P supplying features [16,18]. A larger MBC value represents a stronger P storage capability of soil, resulting in a higher soil P supplying ability. When the intensities of P supplying ability are similar in soils, the MBC value is higher and the storage of effective P is greater, so more effective P is supplied to crops by soil. When P adsorption is comparable among different soils, a larger MBC value indicates a lower energy state of the adsorbed P is, and that the P can be easily used by crops, so the intensity of P supply is relatively small [19]. The MBC value of the NPK treatment was much higher than for the other four treatments, and the CK treatment had substantially higher MBC than the PK and NPKS treatments as shown in table 5. There was no difference among the PK, NPKM and NPKS treatments. The CK had a very large X_m value, and consequently, its MBC value was significantly higher than that of PK and NPKS, but less that for the NPK treatment. Due to similar values for K and X_m in the PK and NPKS treatments, their MBC values did not significantly differ.
- The P sorption saturation (DPS) reflects the ratio between the content of effective P in soil and its maximum *adsorption* [4]: DPS = [available P in soil (Olsen-P)/maximum P adsorption] × 100, where the maximum adsorption of P is calculated by the Langmuir equation. The order of DPS among treatments was NPKM > PK > NPK > NPKS > CK as shown in table 5. Of these, the DPS of NPKS, CK and NPK did not significantly differ, but there were substantial

differences among all other treatments. The variation in *DPS* of all treatments was similar to the trend of Olsen-P content in soil as shown in table 1.

3.2. Desorption characteristics of soil P

3.2.1. Desorption amount and desorption rate of soil P. The desorption of P in soil is usually considered as the reverse process of adsorption, which depends on the reuse of adsorbed P. As exogenous P increased, the desorption amount of P gradually increased in soils with various P contents as shown in figures 3 and 4, indicating that the P desorption amount of soil was related to application of exogenous P. The order of the desorption amount among treatments was CK > NPKM > PK > NPKS > NPK. The desorption rate of P is represented by the ratio between the amount of desorption P and total P adsorbed before desorption, and the desorption rate had a similar pattern to that of desorption amount – it increased with increased exogenous application of P, consistent with previous results [4, 20]. The order of desorption rate among treatments was PK > NPKM > CK > NPKS > NPK, and this pattern was significantly negatively correlated with *K*-value as shown in table 5 – the higher the *K*-value, the smaller was the desorption rate in soil. The *K*-value of NPK was much greater than that of the other four treatments, and thus its desorption rate was relatively low as shown in figure 4.

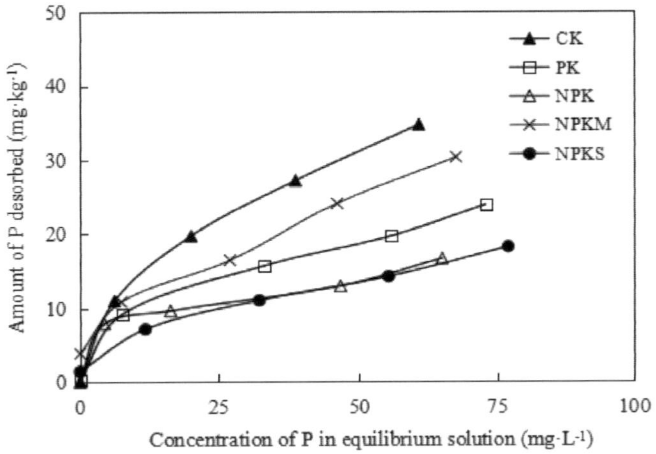

Figure 3. The P isothermal desorption curves in different treatments.

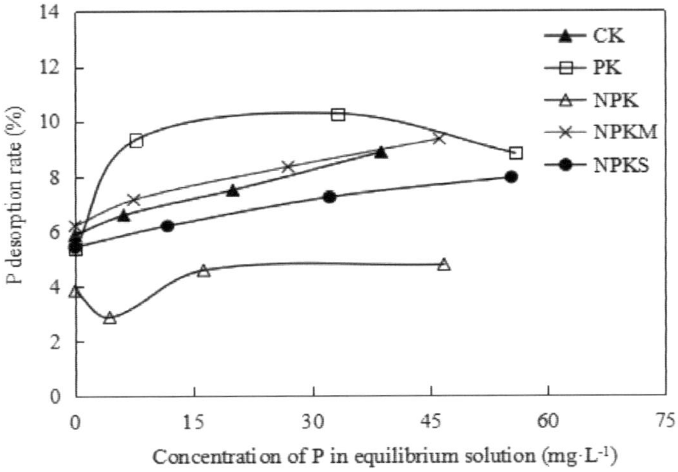

Figure 4. The P isothermal desorption rates in different treatments.

3.2.2. Readily desorbable P in soil (RDP). The order of *RDP* among different treatments was NPKM > NPKS > PK > NPK = CK. Of these, no significant difference was observed among the *RDP* of the PK, NPK and CK treatments, but *RDP* of all other treatments varied substantially. The order of *RDP* values suggested that the chemical fertilizer with organic fertilizer or straw returning significantly improved the content of readily desorbable P in soil, and this effect was much greater than that of chemical fertilization alone. The *RDP* value of NPKM was much higher than that of the other treatments, indicating that chemical with organic fertilization greatly increased the P desorption of soil, and significantly improved the effect of available P in soil.

4. Discussion

Liu *et al* (2003) [11] demonstrated that the content of Olsen-P in soil significantly differed among various treatments (by almost 30 $mg \cdot kg^{-1}$) after 11 years of long-term fertilization experiments in gray desert soil, Xinjiang, China. In the present study, we performed the fertilization experiments for another 11 years, and the difference in the Olsen-P content of various treatments was further magnified with a range reaching 41 $mg \cdot kg^{-1}$. The long-term and different fertilization practices led to differences in the input (fertilization) and output (biomass) of P [21-23], and the adsorption and desorption of soil P changed accordingly. Here, we selected five fertilization treatments with Olsen-P content from low to high, and except for the PK treatment, they represented chronically different fertilization practices of local farmers.

The long-term biomass output without nutrient input in the CK treatment resulted in a lack of P supply in soil, leading to depletion of the soil. Therefore, CK showed a maximum X_m, a minimum *DPS* and no readily desorbable P in the isothermal adsorption experiments. The low content of effective P and high P adsorption resulted in the minimal *DPS* in CK ($P < 0.01$). The NPKM treatment exhibited a relatively large amount of readily desorbable P and a relatively low X_m, and thus the *DPS* of NPKM was much greater than that of other treatments, suggesting that the P level in soil greatly affected the amount and strength of P adsorption, consistent with previous studies [14]. Based on *DPS*, the P requirement for saturated adsorption at 0–200 mm depth for CK, PK, NPK, NPKM and NPKS was 1287, 670, 960, 828 and 691 $kg \cdot ha^{-1}$, respectively. According to the P requirement for saturated P in gray desert soil and annual nutrient inputs, it will theoretically take about 14, 36, 44 and 26 years to saturate the P in the soil for the PK, NPK, NPKM and NPKS treatments, respectively. However, we studied the fertilized topsoil of 0–200 mm and not the whole topsoil of 0–600 mm in the present study, and the sampling time was the late growing stage of winter wheat but not post harvesting, so the actual time to reach P saturation will greatly exceed 14–44 years. This result indicates that the adsorption and desorption of P in gray desert soil is a dynamic and long-term process, and gray desert soil showed an enormous potential for P adsorption.

Compared to the two-season soil with wheat and maize rotation, the *K*-value of gray desert soil (0.18–0.61) was greater than that of Lou (0.07–0.12), Loessial (0.07–0.12) and Loess soils (0.06–0.10) of Shanxi Province and calcareous Chao soil of Hebei Province (0.04–0.10) [5, 24]. This was the main reason for the higher amount and stronger capability of P adsorption in the gray desert soil of Xinjiang with only one season of planting.

5. Conclusion

Within the range of 0–100 $mg \cdot L^{-1}$ for exogenous P concentration, the fit for a Langmuir equation with the isothermal curve of gray desert soil reached a significant level, with $r > 0.95$. The P adsorption, desorption and desorption rate gradually increased with higher exogenous P application in gray desert soil for various levels of Olsen-P, but the adsorption rate was reduced. The soil content of Olsen-P in gray desert soil was significantly positively correlated with the *DPS* of soil – the soil Olsen-P content determined the value of *DPS*, indicating that Olsen-P content was a key factor affecting *DPS*.

The order of soil X_m among different treatments with various Olsen-P accumulations in gray desert soil was CK > NPK ≈ NPKM > PK ≈ NPKS, and the differences among treatments were statistically significant. No substantial difference was observed in *K* among the CK, PK, NPKM and NPKS treatments, but *K* was significantly higher for NPK than the other four treatments. The *MBC* of the NPK treatment was significantly higher compared with the other treatments. The *MBC* of CK was substantially higher than for PK and NPKS, but was similar between CK and NPKM and between PK and NPKS. The *RDP* of NPKM and NPKS treatments was much higher than for treatments with only chemical fertilization, and chemical with organic fertilization exhibited the optimal effect. Therefore, NPKM was the best fertilization practice to elevate the soil P supplying capability.

Acknowledgments
We thank the National Natural Science Foundation of China (41661075 and 41161054), Science and Technology Project Be Supported Xinjiang China (2016E02020) and Xinjiang Key Research and Development Program of China (2016B02017-4) for providing funds for this work. We also thank all the staff associated with long-term soil fertility and fertilizer effects location monitoring network of Xinjiang Grey Desert Soil in China for supplying the historic dataset.

References
[1] Wang B, Liu H, Hao X Y, Wang X H, Sun J S, Li J M and Ma Y B 2016 Agronomic threshold of soil available phosphorus in grey desert soils in Xinjiang, China *IOP Conf. Series: Earth Envir. Sci.* **41** 012010
[2] Wang B, Li J M, Ren Y, Xin J S, Hao X Y, Ma Y B and Ma X W 2015 Validation of a soil phosphorus accumulation model in the wheat-maize rotation production areas of China *Field Crops Res.* **178** 42-8
[3] Wang B, Liu H, Wang X H, Li J M, Ma Y B and Ma X W 2015 Soil phosphorus accumulation model for an arid area of north-western China with 3-year rotation of wheat, maize and cotton *J. Agr. Sci-Cambridge* **153** 1247-56
[4] Chen B L, Sheng J D, Jiang P A and Liu Y G 2010 Study on characteristics of phosphorus adsorption and desorption of cotton field with different soil textures (In Chinese) *Chinese J. Soil Sci.* **41(2)** 303-7
[5] Luo M and Wang X D 2008 Distribution of particles size and their phosphorus adsorption and desorption characteristics of lou soils with different fertility (In Chinese) *Agr. Res. Arid Areas* **26(6)** 106-9
[6] Boparai H K and Sharma K N 2006 Phosphorus adsorption and desorption characteristics of some soils as affected by clay and available phosphorus content *J. Indian Chem. Soc.* **54(1)** 111-4
[7] Zhao X Q and Lu R K 1991 Effect of organic manures on soil phosphorus adsorption (In Chinese) *Acta Pedologica Sinica* **28(1)** 7-13
[8] Han X, Wu D Y and Wang X G 2011 Study on the adsorption characteristics of phosphorus in different soils (In Chinese) *Modern Agr. Sci. Technol.* **20** 285-6
[9] Li C Y, Dang T H, Wang W Z and Liu W Z 2011 Influence of humic acid on the adsorption behavior of phosphorus in agricultural soil (In Chinese) *J. Soil Water Conserv.* **25(3)** 77-82
[10] Yang X Q, Hu T T, Wang X D and Wang F 2006 Effect on PAM on phosphorus adsorption and desorption in lou soil (In Chinese) *J. Soil Water Conserv.* **20(1)** 87-90
[11] Liu H, Wang J L, Tan X X, Zhang X L, Zhao B Q and Zhang F D 2003 The effect of long-term fertilization on phosphor efficiency and modality in grey desert soil (In Chinese) *Xinjiang Agr. Sci.* **40(2)** 111-5
[12] FAO 2006 World reference base for soils resources *World Soil Resource Report No.* **103**. (Rome, Italy)
[13] Bao S D 2007 *Analytical Methods for Soil and Agro-chemistry* (In Chinese) (Beijing: China Agriculture Press)
[14] Zhang H T, Liu J L, Liao W H, Zhang Z X and Hao X Y 2008 Effect of phosphate fertilizer and manure on properties of phosphorus sorption and desorption in soils with different phosphorus levels (In Chinese) *Plant Nutr. Fertilizing Sci.* **14(2)** 284-90

[15] Chen Y D, Liang C H, Wang Y S, Li Y Y, Guo B D and Zhang Y H 2010 Phosphorus adsorption and desorption characteristics of wetland soil under different redox conditions (In Chinese) *Chinese J. Ecol.* **29(4)** 724-9

[16] Zhao Q L, Wang K R and Xie X L 2009 Effects of organic nutrient recycling on phosphorus adsorption-desorption characteristics in a reddish paddy rice system (In Chinese) *Scientia Agr. Sinica* **42(1)** 355-62

[17] Qu H, Zhao B Q, Chen Y H, Liu H, Li X Y and Zhang F D 2004 Effect of long-term fertilization on wheat quality and yield in grey desert soil (In Chinese) *Plant Nutr. Fertilizer Sci.* **10(1)** 12-7

[18] Zhang D, Wei Z M, Li S Q, Wang S P and Xu J G 2005 Effect of bio-organic fertilizers on phosphorus adsorption-desorption (In Chinese) *J. Northeast Agri. Uni.* **36(5)** 571-5

[19] Guo X D, Zhang X Q and Yang L 1997 Phosphorus sorption and desorption properties of agricultural soils in Gansu Province (In Chinese) *Acta Agr. Boreali-Occidentalis Sinica* **6(2)** 7-12

[20] Guo S L, Dang T H, Liu S Z and Hao M D 2005 Changes in characterization of phosphorus sorption in relation to its forms and soil organic carbon (In Chinese) *Plant Nutr. Fertilizer Sci.* **11(1)** 33-9

[21] Ma Y B, Li J M, Li X Y, Tang X, Liang Y C, Huang S M, Wang B R, Liu H and Yang X Y 2009 Phosphorus accumulation and depletion in soils in wheat-maize cropping systems: Model and validation *Field Crops Res.* **110** 207-12

[22] Huang S M, Ma Y B, Bao D, Guo D and Zhang S 2011 Manures behave similar to superphosphate in phosphorus accumulation in long-term field soils *Int. J. Plant Prod.* **5** 135-46

[23] Li J M, Gao J S, Liu J, Xu M G and Ma Y B 2012 Predictive model for phosphorus accumulation in paddy soils with long-term inorganic fertilization *Commun. Soil Sci. Plan.* **43** 1823-32

[24] Xia W J, Liang G Q, Zhou W, Wang H, Wang X B and Sun J W 2008 Adsorption and desorption characteristics of soil phosphorus in calcareous fluvo-aquic soil under long-term fertilization (In Chinese) *Plant Nutr. Fertilizer Sci.* **14(3)** 431-8

Simultaneous determination of 9 heterocyclic aromatic amines in pork products by liquid chromatography coupled with triple quadrupole tandem mass spectrometry

X C Shen[1,2], Y L Zhang[1,2], Y Q Cui[1], L Y Xu[1], X Li[1,2] and J H Qi[1,2,3,4]

[1] College of Food Science and Engineering, Beijing Key Laboratory of Detection and Control of Spoilage Organisms and Pesticide Residues in Agricultural Products, Food Chemistry teaching team, Beijing University of Agriculture, Beijing 102206, China
[2] Pig innovation team, Modern agricultural industry technology system Beijing innovation team, Beijing 100102, China
[3] High Level Talents of Beijing Universities Cross Training "Real Training Plan" Project, Beijing 102206, China

E-mail: 943908793@qq.com

Abstract. Heterocyclic aromatic amines (HAAs) are potent mutagens that formed at high temperature in cooked, protein-rich food. Owing to their frequent intake, an accurate method is essential to access human health risk of HAAs exposure through detecting these compounds in various heat-treated meat products. In this study, a liquid chromatography-electrospray tandem mass spectrometry (LC--ESI-MS/MS) method was developed to perform the determination of 9 mutagenic heterocyclic amines (HAAs) in meat samples with multiple reaction monitoring (MRM) mode. Ultrasound assisted extraction and diatomaceous earth was employed to extract HAAs from food samples, and the analytes were purified and enriched using tandem solid phase extraction, with propyl sulfonic acid coupled to a C18 cartridge. Two parameters, extraction time and eluent, were carefully optimized to improve the extraction and purification efficiency. The LC separation was carried out using a Zorbax SB-C18 (3.5 μm particle size, 2.1×150 mm i.d.) column and optimized some parameters, such as pH, concentration and volume. Under the optimal experimental conditions, recoveries ranged from 52.97% to 97.11% with good quality parameters: limit of detection values between 0.02 and 0.24 ng mL^{-1}, linearity (R^2>0.998), and run-to-run and day-to-day precisions lower than 9.81% achieved. To evaluate the performance of the method in high throughput analysis of complex meat samples, the LC-MS/MS method was applied to the analysis of HAAs in three food samples, and the results demonstrated that the method can be used for the trace determination of HAAs in pork samples.

1. Introduction

Heterocyclic aromatic amines (HAAs) are considered to be carcinogenic, formed during the thermal processing of meat products which are rich in proteins and creatine/creatinine [1]. Recently lots of studies have indicated that many methods of cooking can bring about the generation of chemicals that are harmful to human health [2]. Epidemiological studies suggest an increased cancer risk of the colon, intestine, breast, prostate and pancreas after high intake of cooked meat, especially red meat [3-5]. Red meat contains beef, goat, pork, and lamb [6]. According to the FAO, meat production increased from

96.31 million tons in 2003 to 111.93 million tons at global level total in 2013 (FAO 2013). Besides, in China, pork is the highest consumption and accounted for nearly 65.1% of overall meat production. Therefore, to develop a method to quantity HAAs in cooked pork products is essential. The aim of this study is to establish an accurate and effective analysis method to detect relatively low-level HAAs in pork samples.

Since detected in cooked meat [7], HAAs have attracted much interest. Now, more than 25 types of HAAs have been found [8]. According to structure, HAAs can be divided into thermic HAAs and pyrolytic HAAs. The thermic HAAs, which include IQ, IQx, MeIQ, MeIQx, 4, 8-DiMeIQx, 7, 8-DiMeIQx, PhIP and DMIP, are formed from precursors such as free amino acids, creatine/ creatinine, and hexoses at common processing temperatures in the Maillard reaction [9]. In contrast, pyrolytic HAAs, which contain Harman, Norhaman, AαC, MeAαC, Trp-P-1, Trp-P-2, Glu-P-1 and Glu-P-2, are formed through a pyrolytic reaction at processing temperatures higher than 300℃ [10]. This class of HAAs is not directly mutagenic but can enhance the mutagenicity of other HAAs [11]. There are many factors affecting the generation of HAAs, including type of meat, processing methods, content of precursors [12,13].

Since the complexity of the meat matrix and the trace amounts of HAAs, sample pre-treatment is extremely essential. There are several strategies have been employed in reported methods, such as liquid-liquid extraction [14], extraction with blue cotton [15], solid-phase extraction (SPE) [16]. Based on the SPE method, Gross and GrÜter [17] developed a tandem SPE method (Extrelut-PRS-C18 coupled cartridges), which is the commonly used method, improving the purification efficiency. As to the identification and quantification of HAAs, chromatographic or electrophoretic techniques are commonly carried out, using different methods of detection [18]. These contain liquid chromatography (LC) in conjunction with ultraviolet (UV) [19, 20], fluorescence (FLD) [21], diode array (DAD) [22], or electrochemical (CE) detection [23], gas chromatography–mass spectrometry (GC–MS) [24], liquid chromatography–mass spectrometry (LC–MS) [25-27] and ultra-performance liquid chromatography-mass spectrometry (UPLC-MS) [28]. In the last few decades, the widespread applied for HAAs analysis in food samples is LC-MS, because of its high selectivity and specificity than UV, DAD and FLD, without requirement of analyte derivatization like GC-MS and its lower equipment demanding than UPLC-MS.

In this work, by optimizing an Extrelut-PRS-C18 coupled cartridges method and LC-MS conditions, a LC-ESI-MS/ MS method for the determination of 9 HAAs in pork sample was established. These HAAs are more common and relatively high concentration in the pork samples [29]. To confirm the suitability of the method, pork samples were analyzed to further demonstrate the applicability of the method, which can offer the assessment of human health risk of HAAs exposure.

2. Materials and methods

2.1. Chemicals

The Heterocyclic aromatic amine standards IQ (2-amino-3-methyl-imidazo[4,5-f]quinoline), MeIQ (2-amino-3,4- dimethyl -imidazo [4,5-f] quinoline), MeIQx (2-amino-3,8-dimethyl-imidazo [4,5-f]quinoxaline), 4,8-DiMeIQx (2-amino-3,4,8-trimethyl-imidazo [4,5-f] quinoxaline), TriMeIQx (2-amino-3,4,7,8 -tetramethyl-imidazo [4,5-f] quinoxaline), PhIP (2-amino-1-methyl-6- phenylimidazo [4,5-b]pyridine), AαC (2-amino-9H-pyrido[2,3-b]indole), Trp-P-1 (3-amino-1,4-dimethyl -5H-pyrido[4,3-b]indole), Harman (1-methyl-9H-pyrido[3,4-b]indole), Norharman (9H-pyrido[3,4-b]indole) were obtained from Toronto Research Chemicals Inc. (Toronto, Canada). Stock standards solutions of 100 μg·mL^{-1} in methanol were prepared and stored at 4℃ for further dilutions. Standard mixtures of all HAAs with 4,7,8-TriMeIQx (500 ng· mL^{-1}) as internal standard (IS) at different concentration levels were prepared using methanol to establish the range of linearity and for the calibration curves. Solutions and samples were filters through 0.22 μm nylon filter (Jin teng, China) before analysis.

Methanol, acetonitrile, ammonium formate and formic acid were HPLC grade, provided by J.T.

Baker (Gross-Gerau, Germany) and Aladdin (Shanghai, China) respectively. Water was purified in a Ro water purification system (Heal Force, China). Other chemicals were obtained from Sinopham Chemical Reagent Co., Ltd (Beijing, China). Mobile phases were also filtered with a 0.22 μm nylon filter (Jin teng, China). Extrelut NT20 extraction cartridge and refill material (diatomaceous earth) were from Merk (Darmstadt, Germany). Octadecylsilane (C18,100 mg and 500 mg) SPE columns and propyl sulfonic acid (PRS,500mg) columns were obtained from Agilent Technologies Inc. (America). Coupling pieces and stopcocks were purchased from Agilent (America).

2.2. Instrumentation and MS conditions

Chromatographic separation of HAAs depended on a liquid chromatography system, equipped with a binary pump system from Agilent technologies model series 1200 (Agilent, USA), using Zorbax SB-C18 (3.5 μm, 2.1×150 mm) (Agilent, USA) column. Solvent A: 7 mM formic acid/ ammonium formate pH 3.2; solvent B: acetonitrile. Optimum separation was achieved at a flow rate of 0.3 mL min^{-1}. The gradient elution program was: 0-1min, 5%B; 1-15 min, 30%B; 15-18 min, 60%B; 18-24 min, 60%B; 24-25 min, return to initial conditions; 25-35 min, equilibration of the column. Volume injected was 10 μL. The LC system was coupled to an Agilent 6410 triple quadrupole mass spectrometer equipped with electrospray ionization (ESI) (Agilent, America). The MS instrument was operated in the positive mode and the data was gained in multiple reaction monitoring (MRM) mode. For quantification, 9 analytes were monitored simultaneously. Source working conditions were as follows: spray voltage, 3.5 kV; nebulizer pressure, 30 psi; drying gas flow rate, 10 L·min^{-1} heated capillary temperature, 350℃. Data acquisition was performed by Mass Hunter Workstation Software.

2.3. Sample treatment

Raw pork and commercial pork products were acquired from a local market in Beijing. Raw meat samples were cooked in water at 100℃. And all the samples were fully mixed using a meat grinder and stored at -18℃ before analysis. HAAs were extracted from food samples and purified by the method developed by Gross and GrÜter [17] with some modification [25]. Meat sample (4 g) was homogenized with 40mL of 1mol/L NaOH by Scattering Machine (FSH-2, Changzhou Guohua, China) for 5 min, and extracted by ultrasonic instrument (KQ-100E, Kunshan Shumei, China) for 20 min. Then the cooked food of which quantity equal to 1 g blended with 13 g diatomaceous earth material. The sample was transferred to a vacant Extrelut column coupled to a Bond Elut PRS cartridge (500 mg) conditioned with dichloromethane containing 5% methylbenzene (4 mL). After being extracted from diatomaceous earth, the HAAs were eluted to the PRS cartridge with the usage of dichloromethane containing 5% methylbenzene (75 mL) through solide phase extraction device (12 Position Vacuum Manifold Set, CNW Technologies GmbH, Germany). The PRS cartridge was vacuum dried and rinsed with HCl (0.1 M, 6 mL), methanol /0.1 M HCl (4:6, v/v, 15 mL) and water (2 mL) to wash out the nonpolar HAAs and others impurities. In order to the further purity, ammonia (0.5 mL) and water (20-30 mL) were added to solution containing nonpolar HAAs to dilute to the content of methanol less than 20%. Then added to a Bond Elut C18 (500 mg) cartridge conditioned with methanol (2 mL) and water (10 mL) and washed with water (2 mL), dried through a vacuuming procedure. And with the usage of methanol/ ammonia (9:1, v/v, 1.4 mL), HAAs were eluted to a vial.

Coupled to a Bond Elut C18 (100 mg) cartridge pretreated with methanol (2 mL) and water (10 mL), the PRS cartridge was desorbed with ammonium acetate (0.5 M, 20 mL) regulated at pH8.0. To end, the C18 (100 mg) cartridge was cleaned with water (2 mL), dried through a vacuuming procedure, and then HAAs were eluted to a vial with methanol/ammonia (9:1, v/v, 0.8 mL). The mixture solvent was evaporated by a flow of nitrogen with Sample Concentrator (BF-2000, Ba Fang Century, China) and the residues were re-dissolved with internal standard 4,7,8-TriMeIQx (500 ng·mL-1, 500 μL) in methanol.

To assess the recovery of method added three levels (10 ng·g-1, 20 ng·g-1, 50 ng·g-1) HAAs to food meat samples, and recoveries could be obtained by analyzing the calibration curves.

3. Results and discussion

3.1. Optimization of the chromatographic separation

For the purpose of gaining the best chromatographic separation, buffer pH, concentration and injected volume were studied. In previous work [30], most of the columns for HAAs analysis were C18 or C8 with small particle size (from 3 to 5 μm) providing high efficiency. Taking into account the results obtained, a Zorbax SB-C18 (3.5 μm, 2.1× 150 mm) column was used for the optimization of HAAs separation in this study. In order to protonate the amino groups of the HAAs, the mobile phase pH is required to be lower than the pKa value of the corresponding compound. The pKa values of the HAAs

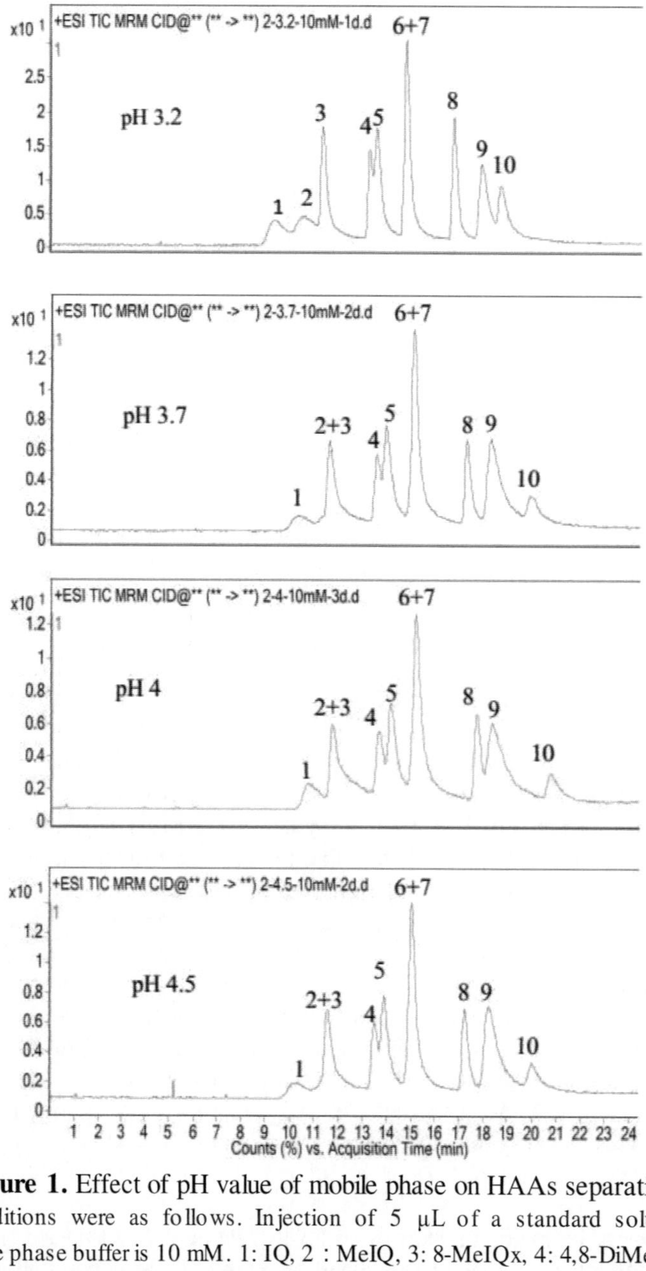

Figure 1. Effect of pH value of mobile phase on HAAs separation.

Chromatographic conditions were as follows. Injection of 5 μL of a standard solution of 100ppb. The concentration of mobile phase buffer is 10 mM. 1: IQ, 2 : MeIQ, 3: 8-MeIQx, 4: 4,8-DiMeIQx, 5: Norharman, 6: Harman, 7: 4,7,8-TriMeI-Qx, 8: PhIP, 9: Trp-p-1, 10: AαC.

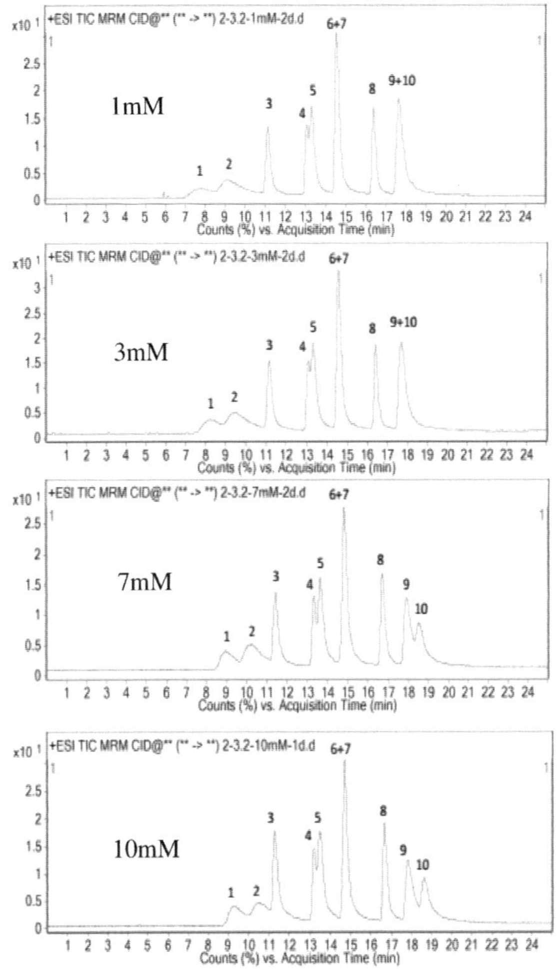

Figure 2. Effect of concentration of mobile phase buffer on HAAs separation.

Chromatographic conditions were as follows. Injection of 5 μL of a standard solution of 100 ppb. pH: 3.2. 1: IQ, 2: MeIQ, 3: 8-MeIQx, 4: 4,8-DiMeIQx, 5: Norharman, 6: Harman, 7: 4,7,8-TriMeIQx, 8: PhIP, 9: Trp-p-1, 10: AαC.

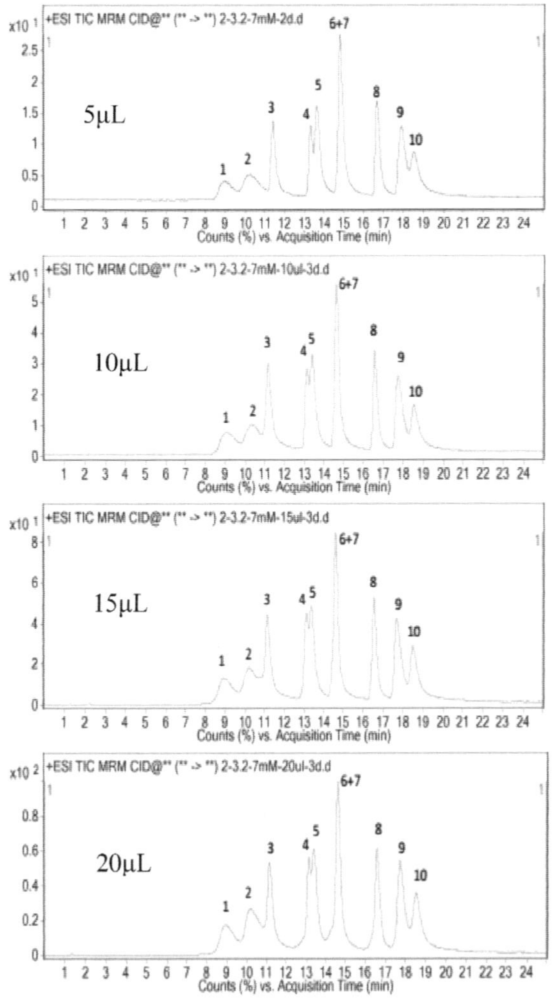

Figure 3. Effect of injected volume of sample on HAAs separation.

Chromatographic conditions were as follows. The concentration of mobile phase buffer is 7 mM, pH 3.2. The concentration of standard solution is 100 ppb. 1: IQ, 2: MeIQ, 3: 8-MeIQx, 4: 4,8-DiMeIQx, 5: Norharman, 6: Harman, 7: 4,7,8-TriMeIQx, 8: PhIP, 9: Trp-p-1, 10: AαC.

studied in this study were mostly in the vicinity of 6, such as IQ, MeIQ, 8-MeIQx, 4,8-DiMeIQx, 4,7,8-TriMeIQx, PhIP, Norharman and Harman, whose values of pKa are 6.4, 5.5, 6.5, 6.0, 6.5, 6.3, and 6.9, respectively [31]. In the positive ion detection mode, the mobile phase pH is generally two units lower than the pKa value of the corresponding HAAs [32]. Therefore, the separation effect of 10 kinds of heterocyclic amines was investigated when the pH value was 3.2, 3.7, 4.0 and 4.5. As shown in figure 1, obviously, a complete separation could not be obtained under any condition. Increasing pH varied from 3.2 to 4.5, the retention time of the HAAs was prolonged, although only slightly variated, and the chromatographic peak of 8-MeIQx (peak 3) and Trp-p-1 (peak 9) widened. Bianchi *et al* [33] also studied the effect of pH on the chromatographic behaviors, as the pH increasing from 2.8 to 4.7, the retention time and peak width increased, which were consistent with this study. Moreover, there was no change in the elution order, the results were consistent with an earlier research [30]. Comparing with previous study [33] (pH 3.0, 3.5, 4.0, 4.75), the elution order of Trp-p-1, PhIP, MeIQ and 8-MeIQx were identical when buffer pH below 4.75, changing at 4.75. Consequently, pH 3.2 was selected since the highest signal responses and the best resolution between MeIQ (peak 2) and 8-MeIQx (peak 3) were obtained.

In terms of concentration of mobile phase, the separation effect of the formic acid-ammonium/formate buffer solution at the concentration of 1, 3, 7, 10 mM was investigated. Figure 2 shows that with concentration lower than 7 mM, Trp-p-1 (peak 9) and AαC (peak 10) co-eluted. With the increasing of the mobile phase concentration, the retention time was increased. When the concentration was 7 mM and 10 mM, the chromatographic peak separation of 10 kinds of heterocyclic amines was relatively good. Compared with 10mM, the retention time was shorter and baseline separation of all the analytes was better at 7mM, so the concentration of 7mM was the best. For the injected volume (5, 10, 15, 20 μL), Figure 3 showed that the concentration had little effect on the separation of HAAs, compared to others, the resolution between Trp-p-1(peak 9) and AαC (peak 10) was worse for 5 μL. As the injected volume varying from 5 to 20 μL, chromatographic peak gained better signal responses. However, as the injection volume increased, the response value did not multiply, 10 μL was established as sufficient.

3.2. Optimization of the MS/MS working conditions

To obtain the best signal responses of all the analytes, we optimized fragmentor of precursor ions and collision energy (CE) of all the product ions in this section. In the positive mode, the precursors ion opted were the protonated molecular ions $[M+H]^+$. Using product scan mode, the product ions were acquired through the process of scanning m/z from 100 to 250 for each HAA at 100ng mL^{-1}. The most abundant product ions were monitored for HAAs quantification, and the second-most abundant product ions were monitored for confirmation of the analytes. Then fragmentor of precursor ions and CE of all the product ions were optimized by MassHunter Optimizer Software, ranged from 70 to 150 eV and between 20 to 50 eV respectively. As shown in table 1, the greatest responses for every precursor and product ion were acquired after optimization.

Table 1. Quantitative and confirmative characteristic ions and optimized MS/MS parameters of HAAs.

Analyte	Precursor ion (m/z)	Fragmentor (eV)	Quantitation		Confirmation	
			Product ion (m/z)	Collision energy (eV)	Product ion (m/z)	Collision energy (eV)
IQ	199.1	80	184.3	27	157.0	40
MeIQ	213.0	115	198.1	27	197.0	39
8-MeIQx	214.2	115	199.1	27	131.2	40
4,8-DiMeIQx	228.1	115	213.1	27	186.9	23
4,7,8-TriMeIQx	242.0	115	226.8	27	201.2	23
PhIP	225.2	115	210.1	31	182.8	39
Norharman	169.1	140	114.8	31	167.8	35
Harman	183.0	130	114.9	27	168.1	39
Trp-P-1	212.1	135	195.0	23	168.0	31
AαC	184.0	130	167.1	23	139.8	35

The results agreed with previous studies [34, 35] for the majority ion of all the analytes. For polar compounds, such as IQ, MeIQ, 8-MeIQx, 4,8-DiMeIQx, 4,7,8-TriMeIQx and PhIP, the product ions $[M+H-CH_3]^+$ that provided the maximal intensity resulted from the loss of the methyl group, while for nonpolar HAAs, such as Trp-p-1 and AαC, the ammonia-loss ion, $[M+H-NH_3]^+$, was the predominant fragment. However, for the qualitative ions of 8-MeIQx, 4, 8-DiMeIQx and Norharman, there were some differences between this study and others [18, 25, 36]. The reasons of these results may relate to the difference of buffer pH and concentration [25], mass spectrometer [36], ion source [37] and monitoring mode [18]. Consequently, as shown in figure 4, all compounds were baseline-separated, also, a more accurate quantification could be behaved in MRM mode after optimization.

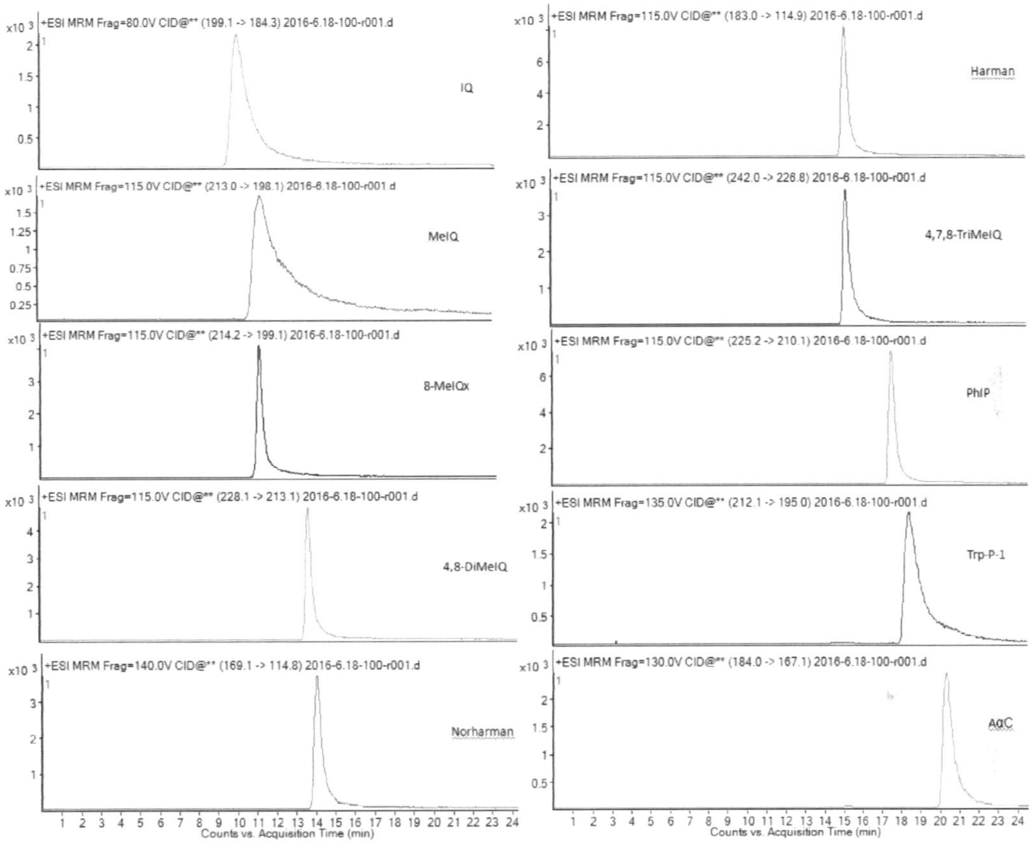

Figure 4. The LC-MS/MS chromatograms of HAAs at standard solution of 100ppb in MRM.

3.3. Optimization of sample treatment

3.3.1. Ultrasound assisted extraction time. The extracting efficiency were assessed on the recovery that achieved. This procedure influenced by many parameters, such as the concentration of alkaline, the selection of solvents and the time of extraction [37, 38]. The extraction time and solvents were studied. Extraction time is the necessary time to release the HAAs from matrix completely by ways of the treatment of initial alkaline. Several periods (0, 5 and 20 min) of Ultrasound assisted extraction were studied after 5min homogenized with the solution of sodium hydroxide. Figure 5 shows the results. With the prolongation of ultrasound time, the recoveries of IQ, MeIQ, 8-MeIQx, Harman and Trp-P-1 were increased. However, for 4,8-DiMeIQx, Norharman, AαC and PhIP, the recoveries were lowest at 5 min , while the recovery of PhIP was highest at the time of non-ultrasonography. Except for IQ and MeIQ, there was no significant change on the recoveries of other HAAs between groups (P> 0.05). As to IQ and MeIQ, compared with other groups, their recoveries had significant difference (P <0.05) at 20 min. In short, when extraction time was 20 min, the overall recovery of heterocyclic amines was better, so 20min was fixed as the optimal ultrasonic extraction time.

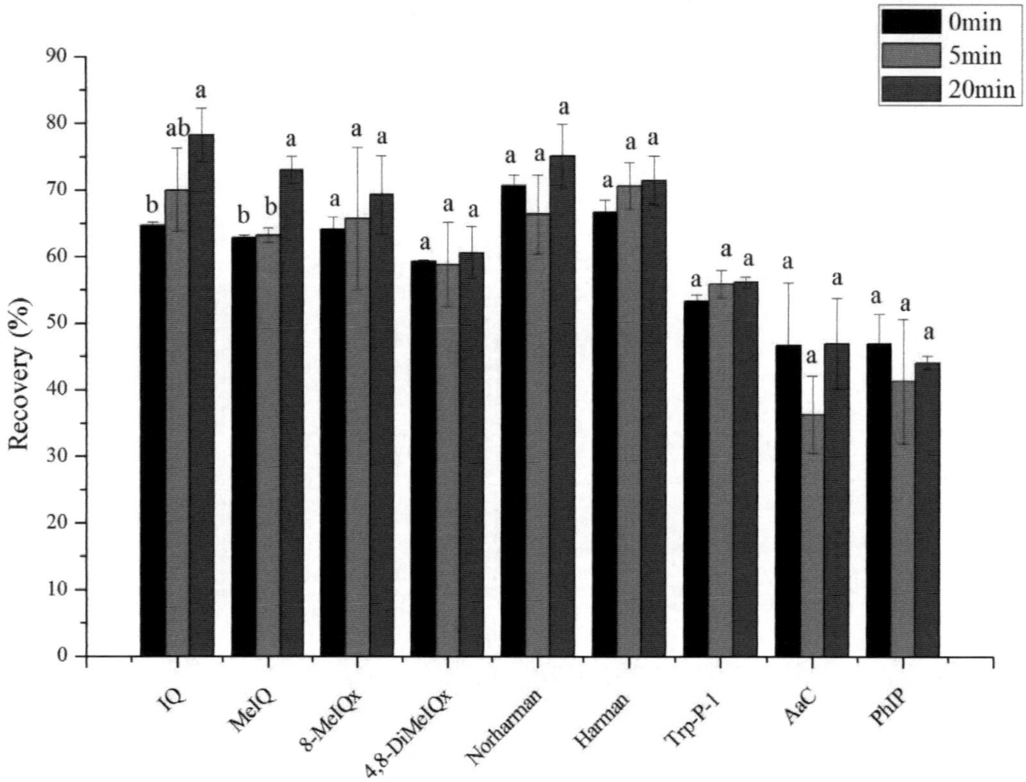

Figure 5. Effects of extraction time on the recoveries of HAAs.

3.3.2. Selection of solvents. For the extracted solvent of heterocyclic amines, the more widely used are dichloromethane and ethyl acetate. Related report [39] showed that owing to the presence of benzene ring in the PhIP structure, according to the similar compatibility principle, the addition of 5% methylbenzene to the solvent improved the recovery of PhIP, so this study selected dichloromethane, ethyl acetate, dichloromethane/ methylbenzene (v/v = 95/5) and ethyl acetate/ methylbenzene (v/v = 95/5) as extraction solvent to estimate the extracting efficiency of heterocyclic amines. As can be seen in figure 6, dichloromethane/ methylbenzene (v/v = 95/5) gained the highest recoveries for all HAAs, ranging from 50.92%-87.12%. However, no significant difference in the recoveries of IQ and MeIQ was seen between groups (P> 0.05). Compared with the experimental group without adding 5% methylbenzene, the addition of group could significantly improve the recovery of AαC (P <0.05). Whether 5% methylbenzene was added to dichloromethane or not, for the remaining heterocyclic amines, their recoveries were higher than that of ethyl acetate and ethyl acetate supplemented with 5% methylbenzene, significantly (P <0.05). Compared with the group of dichloromethane, the recovery of Trp-P-1 and AαC was significantly improved by adding 5% methylbenzene in dichloromethane as extraction solvent (P <0.05). Thus, dichloromethane/ methylbenzene (v/v = 95/5) was opted as the optimum solvent of extraction.

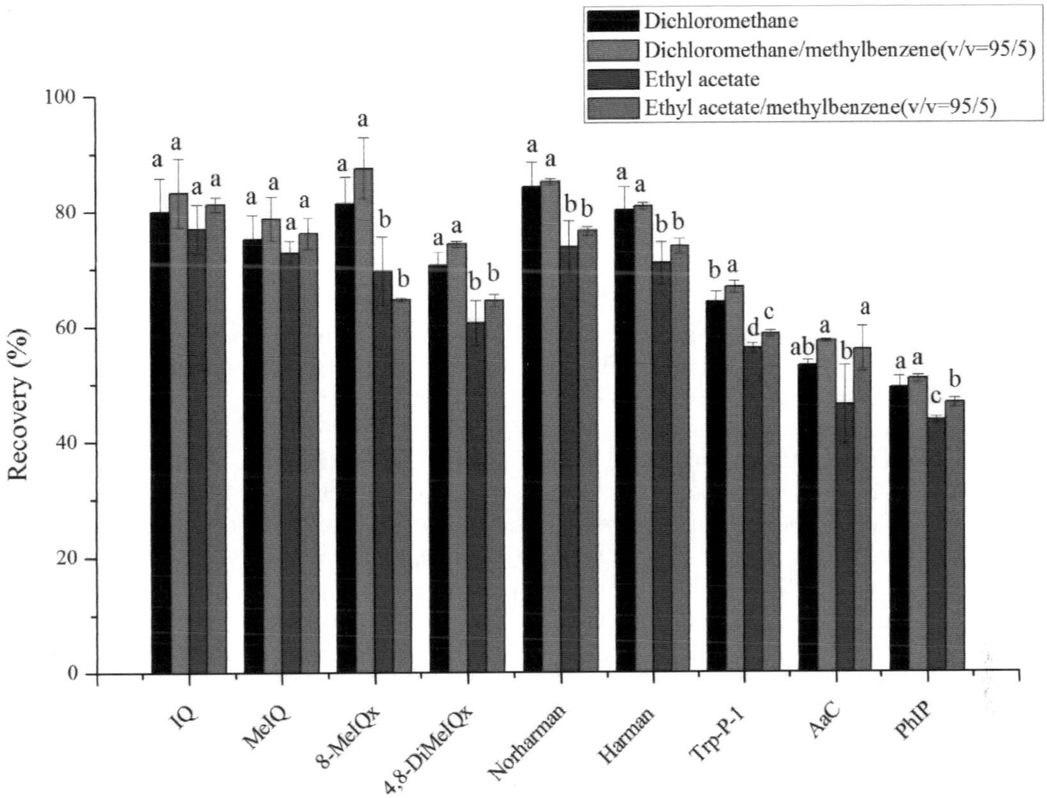

Figure 6. Effects of extraction solvents on the recoveries of HAAs.

Table 2. Quality parameters of the LC-ESI-MS/MS method.

Compound	Regression line			LODs (ng g^{-1})	LOQs (ng g^{-1})	RSD% (n=6)	
	Slope	Intercept	Correlation coefficient			Spiked level 20 ng/g	Spiked level 10 ng/g
IQ	7.3038	-0.0488	0.9996	0.03	0.11	7.80	7.94
MeIQ	7.7372	-0.0840	0.9995	0.02	0.05	4.56	8.68
8-MeIQx	4.9773	0.0100	0.9998	0.17	0.55	1.62	9.81
4,8-DiMeIQx	5.4622	0.0218	0.9992	0.18	0.59	2.15	8.61
Norharman	8.4165	0.0704	0.9994	0.03	0.08	3.53	6.14
Harman	5.5082	-0.0162	0.9986	0.24	0.80	1.92	8.21
Trp-P-1	10.0312	-0.0084	0.9985	0.09	0.30	6.08	7.17
AaC	6.2248	-0.0608	0.9992	0.20	0.65	6.39	7.27
PhIP	4.1632	0.0126	0.9999	0.20	0.65	2.90	7.21

3.4. Evaluation of the LC-ESI-MS/MS method

To assess the LC-MS/MS method's performance, quality parameters such as linearity, the limits of

IOP Conf. Series: Earth and Environmental Science **77** (2017) 012021 doi:10.1088/1755-1315/77/1/012021

detection (LODs), the limits of qualification (LOQs), repeatablility (run-to-run precision) and reproducibility (day-to-day precision) were studied. Calibration carves were selected relative concentration ratio of each heterocyclic amine to 4,7,8-TriMeIQx (internal standard) as the abscissa, the relative peak area ratio as the ordinate. The regression equation and the correlation coefficient were obtained. The LODs, based on a signal-to-noise ratio of 3, and LOQs, depending on a signal-to-noise ratio of 10 were calculated by injecting 10 μL of diluted HAAs standard solutions. Table 2 shows the results. PhIP, Trp-P-1 and AαC had good linearity in the concentration range of 10-500 ng/mL under the optimized chromatographic and mass spectrometric conditions, and the other six in the range of 5-500 ng/mL. Correlation coefficients of all curves were over 0.998. The range of LODs was from 0.02 to 0.24 $ng \cdot g^{-1} ng\ g^{-1}$and LOQs ranged from 0.05 to 0.80 $ng\ g^{-1}$. Run-to-run precision based on concentration was structured via six consecutive injections of a standard solution at spiked levels of 10ng/g and 20 ng/g. To access reproducibility, repeated the above step with the same solution of both two levels over three consecutive days. And accuracy was established by 6 replicate extractions performed with same solution.

High repeatability, reproducibility and accuracy were gained for each compound with relative standard deviations lower than 12.51% in all cases. Moreover, the recoveries of different spiked levels of pork samples were studied, and the results were shown in Table 3. For all HAAs, the recoveries were higher than 50.33%, and the average recoveries ranged from 52.97% to 97.11%. All of results confirmed that the method is successful in providing reliable values of repeatability and reproducibility essential to an exact HAAs analysis.

Table 3. The recovery (R, %) of different spiked levels of pork samples.

Compound	Spiked levels ($ng \cdot g^{-1}$)			Average recovery (%)
	10	20	50	
IQ	78.05 ± 3.78	82.56 ± 2.27	90.78 ± 2.21	83.80
MeIQ	64.98 ± 5.16	59.51 ± 1.84	68.49 ± 2.19	64.33
8-MeIQx	77.25 ± 2.18	77.32 ± 1.91	78.58 ± 2.43	77.72
4,8-DiMeIQx	83.54 ± 6.17	85.97 ± 3.80	85.73 ± 4.02	85.08
Norharman	77.03 ± 7.31	90.44 ± 9.79	85.36 ± 10.07	84.28
Harman	95.77 ± 1.38	106.93 ± 11.02	88.64 ± 8.31	97.11
Trp-P-1	52.96 ± 1.44	55.93 ± 10.31	56.05 ± 9.38	54.98
AaC	50.33 ± 4.55	53.45 ± 5.54	54.13 ± 2.47	52.97
PhIP	68.11 ± 5.41	67.08 ± 4.22	71.02 ± 4.38	68.74

Table 4. HAAs found in three food samples.

HAAs	Concentration ± SD[a] ($ng \cdot g^{-1}$)		
	Pork loss	Roast bacon	Roast fillet
IQ	0.98±0.01	4.96±0.26	nd
MeIQ	0.54±0.01	0.66±0.02	0.59±0.02
8-MeIQx	0.05±0.01	0.32±0.02	0.12±0.03
4,8-DiMeIQx	0.02±0.02	nd	0.13±0.05
Norharman	3.37±0.79	7.08±0.37	3.66±0.10
Harman	11.81±0.51	2.79±0.11	3.81±0.15
Trp-P-1	nd	nd	nd
AaC	nd	nd	0.02±0.01

PhIP	0.09±0.02	0.10±0.01	0.12±0.01

[a] Standard deviation (n=3);
nd: not detected.

For the purpose of evaluating the applicability of the LC-MS method for the analysis of food samples, three pork samples were analyzed (pork loss, roast bacon and roast fillet). Each sample analysis was repeated three times. Table 4 shows the consequences. Low level of HAAs was in most pork products. Among the HAAs detected, Norharman and Harman were found at a high content of all the samples, and the highest concentration of 7.08 and 11.81 $ng{\cdot}g^{-1}$ in Roast bacon and food loss respectively.

4. Conclusions

By optimizing the pretreatment conditions, chromatographic conditions and mass spectrometry conditions, the method of simultaneous determination of 9 kinds of heterocyclic amines in pork through solid phase extraction and liquid chromatography tandem triple quadrupole mass spectrometry was established. The samples were extracted with dichloromethane (containing 5% methylbenzene), purified by sodium propane sulfonate (PRS) column and C18 solid phase extraction column, and analyzed through column chromatography Zorbax SB-C18 (3.5 μm, 150 mm) at optimum condition. The results showed that the correlation coefficient of heterocyclic amine compounds was more than 0.998, the detection limit was 0.02-0.24 ng/mL, the limit of quantification was 0.05-0.65 ng/mL. The average recoveries were 52.97% -97.11%, and RSD were lower than 9.81%. Finally, the method was validated by analyzing three food meat products. The results of the quality parameters achieved, confirm that they are sufficient to propose it as a method for the reliable analysis of HAAs on food samples.

Acknowledgments
The authors are grateful for the financial support from Beijing Innovation Consortium of Swine Research System (Project no. 5075237051/001), and Professor Z Y Wang for technical support.

References

[1] Pais P, Salmon C P, Knize M G and Felton J S 1999 Formation of mutagenic/carcinogenic heterocyclic amines in dry-heated model systems, meats, and meat drippings *J. Agri. & Food Chem.* **47** 1098
[2] Shabbir M A, Raza A, Anjum F M, Khan M R and Suleria H A 2015 Effect of thermal treatment on meat proteins with special reference to heterocyclic aromatic amines (HAAs) *Cri. Reviews in Food Sci. Nutri.* **55** 82-93
[3] Jägerstad M and Skog K 2005 Genotoxicity of heat-processed foods *Mut. Res./Funda. Molec. Mecha. Muta.* **574** 156-72
[4] Sugimura T, Wakabayashi K, Nakagama H and Nagao M 2004 Heterocyclic amines: Mutagens/ carcinogens produced during cooking of meat and fish *Cancer Sci.* **95** 290-9
[5] Ferguson L R 2010 Meat and cancer *Meat Sci.* **84** 308-13
[6] Aaslyng M D, Duedahlolesen L, Jensen K and Meinert L 2013 Content of heterocyclic amines and polycyclic aromatic hydrocarbons in pork, beef and chicken barbecued at home by Danish consumers *Meat Sci.* **93** 85
[7] Lijinsky W and Shubik P 1964 Benzo(a)pyrene and other polynuclear hydrocarbons in charcoal-broiled *Meat Sci.* **145** 53-5
[8] Oz F and Zikirov E 2015 The effects of sous-vide cooking method on the formation ofheterocyclic aromatic amines in beef chops *Leb.-Wis.t und-Technol.* **64** 120-5
[9] Alaejos M S and Afonso A M 2011 Factors that affect the content of heterocyclic aromatic amines in foods *Compre. Reviews Food Sci. Food Saf.* **10** 52-108
[10] Skog K, Solyakov A and Jägerstad M 2000 Effects of heating conditions and additives on the formation of heterocyclic amines with reference to amino-carbolines in a meat juice model

IOP Conf. Series: Earth and Environmental Science **77** (2017) 012021 doi:10.1088/1755-1315/77/1/012021

system *Food Chem.* **68** 299-308

[11] Skog K and Solyakov A 2002 Heterocyclic amines in poultry products: A literature review *Food & Chem. Tox. An Inter. J. Pub. Bri. Ind. Biol. Res. Associ.* **40** 1213-21

[12] Felton J S, Malfatti M A, Knize M G, Salmon C P, Hopmans E C and Wu R W 1997 Health risks of heterocyclic amines *Mut. Res./Funda. Molec. Mech. Mut.* **376** 37-41

[13] Oz F, Kaban G and Kaya M 2010 Effects of cooking methods and levels on formation of heterocyclic aromatic amines in chicken and fish with Oasis extraction method *LWT - Food Sci. Tech.* **43** 1345-50

[14] GuY S, Kim I S, Ahn J K, Park D C, Dong M Y, Ji C I and Kim S B 2002 Mutagenic and carcinogenic heterocyclic amines as affected by muscle types/skin and cooking in pan-roasted mackerel *Mut. Res.* **515** 189-195

[15] Skog K 2004 Blue cotton, Blue Rayon and Blue Chitin in the analysis of heterocyclic aromatic amines--a review *J. Chrom. B Ana. Tech. in the Biomed. Life Sci.* **802** 39-44

[16] Prabhu S, Lee M J, Hu W Y, Winnik B, Yang I, Buckley B and Hong J Y 2001 Determination of 2-Amino-1-methyl-6-phenylimidazo[4,5-b] pyridine (PhIP) and its metabolite 2-Hydroxyamino-PhIP by liquid chromatography/electrospray ionization–ion trap mass spectrometry. *Anal. Biochem.* **298** 306-13

[17] Gross G A, Grüter A and Heyland S 1992 Optimization of the sensitivity of high-performance liquid chromatography in the detection of heterocyclic aromatic amine mutagens *Food Chem. Tox.* **30** 491-8

[18] Zhang Y, Lin C, Fang G, Mei J, Wang X and Wang S 2012 Tandem solid phase extraction coupled to LC-ESI–MS/MS for the accurate simultaneous determination of five heterocyclic aromatic amines in processed meat products *European Food Res. Tech.* **234** 197-205

[19] Zhang Q, Li G and Xiao X 2015 Acrylamide-modified graphene for online micro-solid-phase extraction coupled to high-performance liquid chromatography for sensitive analysis of heterocyclic amines in food samples *Talanta* **131** 127

[20] Raza A, Shabbir M A, Khan M I, Suleria H A R and Sultan S 2015 Effect of thermal treatments on the formation of heterocyclic aromatic amines in various meats. *J. Food Proc. Pres.* **39** 376-83

[21] Viegas O, Moreira P S and Ferreira I M 2015 Influence of beer marinades on reduction of carcinogenic heterocyclic aromatic amines in charcoal-grilled pork meat. *Food Addit. & Contam. Part A Chem. Anal. Contr. Expos. Risk Assess.* **32** 315-23

[22] Oz F, Kızıl M, Zaman A and Turhan S 2016 The effects of direct addition of low and medium molecular weight chitosan on the formation of heterocyclic aromatic amines in beef chop *LWT - Food Sci. Tech.* **65** 861-7

[23] Gerbl U, Cichna M, Zsivkovits M, Knasmuller S and Sontag G 2004 Determination of heterocyclic aromatic amines in beef extract, cooked meat and rat urine by liquid chromatography with coulometric electrode array detection. *J. Chrom. B Ana. Tech. Biomed. & Life Sci.* **802** 107-13

[24] Janoszka B, Błaszczyk U, Warzecha L, Luks-Betlej K and Strózyk M 2003 The analysis of aminoazaarenes as their derivatives with GC-MS method in the heat-processed meat samples *Chemia Anal.* **48** 707-21

[25] Khan M R 2015 Influence of food condiments on the formation of carcinogenic heterocyclic amines in cooked chicken and determination by LC-MS/MS *Food Addit. Contam. Part A Chem. Anal. Control Expo. Risk Assess* **32** 307-14

[26] Hasnol N D, Jinap S and Sanny M 2014 Effect of different types of sugars in a marinating formulation on the formation of heterocyclic amines in grilled chicken *Food Chem.* **145** 514-21

[27] Szterk A 2013 Chemical state of heterocyclic aromatic amines in grilled beef: evaluation by in vitro digestion model and comparison of alkaline hydrolysis and organic solvent for extraction *Food Chem. Toxicol.* **62** 653-60

[28] Turesky R J, White K K, Wilkens L R and Marchand L L 2015 Caffeine cytochrome P450 1A2 metabolic phenotype does not predict the metabolism of heterocyclic aromatic amines in humans *Chem. Res. Toxicol.* **28** 1603

[29] Alaejos M S and Afonso A M 2011 Factors that affect the content of heterocyclic aromatic amines in foods. *Comprehens. Rev. in Food Sci. & Food Saf.* **10** 52-108

[30] Barceló-Barrachina E, Moyano E, Puignou L and Galceran M T 2004 Evaluation of reversed-phase columns for the analysis of heterocyclic aromatic amines by liquid chromatography-electrospray mass spectrometry *J. Chrom. B* **802** 45-59

[31] Sanz Alaejos M, Ayala JH, Gonzalez V and Afonso A M 2008 Analytical methods applied to the determination of heterocyclic aromatic amines in foods *J. Chromatogr. B. Analyt. Technol. Biomed. Life Sci.* **862** 15-42

[32] Barcelo-Barrachina E, Moyano E, Galceran M T, Lliberia J L, Bago B and Cortes M A 2006 Ultra-performance liquid chromatography-tandem mass spectrometry for the analysis of heterocyclic amines in food *J. Chromatogr. A* **1125** 195-203

[33] Bianchi F, Careri M, Corradini C, Elviri L, Mangia A and Zagnoni I 2005 Investigation of the separation of heterocyclic aromatic amines by reversed phase ion-pair liquid chromatography coupled with tandem mass spectrometry: the role of ion pair reagents on LC-MS/MS sensitivity *J. Chromatogr. B Analyt. Tech. Biomed. Life Sci.* **825** 193-200

[34] Busquets R, Mitjans D, Puignou L and Galceran M T 2008 Quantification of heterocyclic amines from thermally processed meats selected from a small-scale population-based study *Molecul. Nutrit. Food Res.* **52** 1408-1420

[35] Shin J H, Na Y C, Chung J H, Gorinstein S and Ahn Y G 2014 Quantitative analysis of heterocyclic amines in urine by liquid chromatography coupled with tandem mass spectrometry *Anal. Biochem.* **447** 169-76

[36] Barceló-Barrachina E, Moyano E, Puignou L and Galceran M T 2004 Evaluation of different liquid chromatography-electrospray mass spectrometry systems for the analysis of heterocyclic amines *J. Chrom. A* **1023** 67-78

[37] Toribio F, Moyano E, Puignou L and Galceran M T 2002 Ion-trap tandem mass spectrometry for the determination of heterocyclic amines in food J. *Chroma. A* **948** 267-81

[38] Khan M R, Busquets R, Santos F J and Puignou L 2008 New method for the analysis of heterocyclic amines in meat extracts using pressurised liquid extraction and liquid chromatography-tandem mass spectrometry *J. Chrom. A* **1194** 155-60

[39] Liao G Z, Wang G Y, Xu X L and Zhou G H 2010 Effect of cooking methods on the formation of heterocyclic aromatic amines in chicken and duck breast *Meat Sci.* **85** 149-54

Institute of Physics
Dirac House, Temple Back
Bristol BS1 6BE UK

ISSN: 1755-1307
ISBN 978-1-5108-4556-5

CURRAN ASSOCIATES INC.
proceedings.com

XVIII Brazilian Colloquium on Orbital Dynamics (CBDO 2016)

Journal of Physics: Conference Series Volume 911

Aguas de Lindoia, Brazil
28 November - 2 December 2016

IOP
Institute of Physics